INDIVIDUAL DEVELOPMENT AND EVOLUTION

INDIVIDUAL DEVELOPMENT AND EVOLUTION

The Genesis of Novel Behavior

GILBERT GOTTLIEB

New York Oxford
OXFORD UNIVERSITY PRESS
1992

Oxford University Press

Oxford New York Toronto
Delhi Bombay Calcutta Madras Karachi
Petaling Jaya Singapore Hong Kong Tokyo
Nairobi Dar es Salaam Cape Town
Melbourne Auckland

and associated companies in
Berlin Ibadan

Published by Oxford University Press, Inc.,
200 Madison Avenue, New York, New York 10016

Library of Congress Cataloging-in-Publication Data
Gottlieb, Gilbert, 1929–
Individual development and evolution :
The genesis of novel behavior
Gilbert Gottlieb.
p. cm.
Includes bibliographical references and index.
ISBN 0-19-506893-9
1. Nature and nurture. 2. Behavior evolution.
I. Title. QH438.5.G68 1992
575—dc20 91-6714

2 4 6 8 9 7 5 3 1

Printed in the United States of America
on acid-free paper

This book is dedicated to my wife,
Nora Lee Willis Gottlieb

Preface

This work is intended to portray the interrelationship of heredity, individual development, and the evolution of species in a way that can be understood by nonspecialists. The ultimate aim of dissolving the nature–nurture dichotomy will be achieved only through the establishment of a fully developmental theory of the phenotype from gene to organism. In striving to offer a straightforward historical exposition of the complex topic of nature and nurture, I have opted to tell the story through a central cast of characters beginning with Lamarck in 1809 and ending with a synthesis of my own making that depicts how extragenetic behavioral changes in individual development could be the first stages in the pathway leading to evolutionary change (genetic change). On the way to that goal, I first describe relevant conceptual aspects of genetics, embryological development, and evolutionary biology in a nontechnical and, I hope, interesting and accurate way for students and colleagues in the behavioral and social sciences. The book presents a highly selective review of the main ways of thinking about heredity, development, and evolution since 1809 as a prelude to the description of a developmental theory of the phenotype in which behavioral change leads eventually to evolutionary change.

In essence, I have three goals in mind in presenting the various ways of thinking about heredity, individual development, and evolution. First, I want to establish the relevance of individual development to the evolution of species. Second, I want to describe what I believe to be the most appropriate way to think about or conceptualize heredity in relation to individual development. The third and most difficult aim is to show that this somewhat unorthodox manner of conceptualizing heredity and individual development gives rise to a new way to think about the behavioral pathway leading to evolution. In sum, I see the present work not only as a contribution toward the possible dissolution of the nature–nurture dichotomy but also as a contribution to evolutionary theory. Individual development has not been included in the neo-Darwinian theory of evolution—it is perhaps the single most important topic yet to be incorporated in the "unfinished synthesis."

My approach is to portray these issues from a current scientific perspective, using certain historical figures to exhibit how certain ways of thinking about development and evolution have become entrenched, thereby setting the stage for the presentation of a new theory of "behavioral neophenogenesis" (how extra- or supragenetic developmental change produced in individuals can eventually lead to genetic evolutionary change). Consequently, the method is not one of a scholarly historian of science but of a working scientist groping publicly for solutions to recalcitrant conceptual issues of the day.

The book grew out of an invited interdisciplinary course of lectures for advanced undergraduates and graduate students delivered at the University of Colorado at Boulder in the summer of 1985. I continued to strive to master these issues in several graduate seminars taught at the University of North Carolina at Greensboro. I am especially indebted to my close colleagues Timothy Johnston, Robert Lickliter, Lynda Uphouse, David Miller, and Robert Cairns in helping me to clarify the relationship of individual development to evolution.

Preparation of the monograph was aided by a research leave from the University of North Carolina at Greensboro during the fall of 1986. I am very grateful for career-long research support from the North Carolina Department of Mental Health (1959–1982), the National Institute of Child Health and Human Development (1962–1985), the National Science Foundation (1985–1988), and the National Institute of Mental Health (1989–1992). Ramona Rodriguiz gave unstinting bibliographic and other assistance during the final phase of the project.

Greensboro, N.C.
December 1990

G. G.

Acknowledgments

The author gratefully acknowledges permission to reproduce material from the following sources: National Library of Medicine, Bethesda, Maryland (Figures 1–1, 2–1, 3–1, 4–1, 6–1, 7–1, 7–2, 8–4, 10–3, 11–1, 11–4 [C. L. Morgan and J. M. Baldwin]); The Natural History Museum London (Figure 3–2); Professor H. J. Jerison and Macmillan Magazines Ltd. (Figure 3–3); D. Appleton, New York (Figure 4–3); Columbia University press (Figure 5–1); New York Academy of Medicine (Figure 7–3); Archiv für Kunst und Geschichte, Berlin (Figure 7–4); Professor Patrick Bateson and the Department of Zoology, University of Cambridge (Figure 8–1); Ferdinand Hamburger, Jr., Archives of the Johns Hopkins University (Figure 8–2); British Association for the Advancement of Science (Figure 9–1); Academic Press (Figure 9–2); Cambridge University Press (Figure 9–4); Oxford University Press (Figure 9–5); Birkhäuser Press (Figure 9–6); Joan Fisher Box (Figure 10–1); Society for the Study of Evolution (Figure 10–4); University of Chicago Press (Figure 10–5); Professor Ernst Mayr (Figure 11–2); Dr. Olga Schmalhausen, Institute of Developmental Biology, USSR Academy of Sciences (Figure 11–3); Department of Library Services, American Museum of Natural History (Figure 11–4: H. F. Osborn); Professor Aubrey Manning and Department of Genetics, University of Edinburgh (Figure 11–5); Professor A. Gorbman and John Wiley (Figure 12–1); American Physical Society and Mrs. Paul Weiss (Figure 12–2); Professor E. B. Lewis, California Institute of Technology (Figure 12–3); Prentice-Hall, New York (Figure 12–4); Dr. J. M. Tanner, Castlemead Publications, Herts, England, and Harvard University Press (Figure 12–5); Professor V. B. Wigglesworth and Weidenfeld and Nicholson (Figure 12–6); Professors Rudolf A. Raff and Thomas C. Kaufman (Figure 12–7); Canadian Psychological Association (Figures 14–2a, 14–2b; Macmillan Magazines Ltd. (Figure 14–5); Professor H. J. Jerison and University of Chicago Press (Figure 14–6); and E. J. Brill, Leiden, Netherlands (Figure 14–7).

Requests for permission to reproduce figures from the following actual or potential copyright holders were to no avail: Figure 8–3 (drawing of Karl Pearson), Charles Griffin and Edward Arnold Publishers, London; Figure 9–3 (photograph of Sir Gavin de Beer by Elliott & Fry), Thomas Nelson and Sons Ltd., and the heirs of Sir Gavin; Figure 10–2 (photograph of Sewall Wright by Hildegard Adler), heirs of Sewall Wright.

I am most grateful to the Creative Division of the Learning Resources Center, University of North Carolina at Greensboro, for the patient and careful preparation of the art work for virtually the entire book.

Finally, I thank Grace Martin and Linda Brown for their cheerful perseverance at the keyboard over the course of several years as the final draft of the manuscript slowly took shape.

Contents

INDIVIDUAL DEVELOPMENT AND EVOLUTION

Conceptions of Development: Preformation and Epigenesis

Preformation and Epigenesis

From the time of Aristotle there have been two competing views about the nature of the process by which individual organisms develop from fertilization to adulthood (= ontogeny). One of these, called *preformation,* held that a very tiny version of the complete individual is prepackaged in the egg—that is, all the parts and organs are preformed from the outset and individual development consists merely of the growth of these preexisting, fully formed parts until they reach their adult size. The alternative view, the one held by Aristotle and almost every other person who took the trouble to make observations on the growing embryo and fetus, holds that individual development takes place by transformations that bring each part and organ of the body into existence in a series of successive stages. This emergent view, called *epigenesis,* is the one we hold today. It signifies that individual development includes differentiation as well as growth, the hallmark of which is the progressive change from an initial relatively homogeneous state to a later highly heterogeneous state. We still do not completely understand all the essential mechanisms whereby epigenesis takes place, but we do know that it is correct to say that individual development—psychologically, behaviorally, physiologically, and anatomically—is epigenetic and not preformative.[1]

Preformation: Ovists and Animalculists

For our historical understanding it is of some value to describe briefly the two main versions of preformation. Since according to this view the organism was preformed in miniature from the outset, the preformed organism was believed by some to lie dormant in the ovary of the female until development was started by fertilization. This view was held by the *ovists.* To other thinkers, the preformed organism resided in the semen of the male and development was unleashed through sexual union with the female. These were the *animalculists.*

Many of the preformationists, whether ovists or animalculists, tended to be of a religious persuasion. In that case they saw the whole of humankind having been originally stored in the ovaries of Eve if they were ovists or in the semen of Adam if they were animalculists. Based upon what was known about the population of the world in the 1700s—at the time of the height of the argument between the ovists and animalculists—Albrecht Haller, the learned physiologist at the University of Göttingen, calculated that God, in the sixth day of his work, created and encased in the ovary of Eve 200 thousand million fully formed human miniatures. Haller was a very committed ovist.

The sad fact about this controversy was that the very best evidence to date for epigenesis was at hand when Haller made his pronouncement for preformation: "There is no coming into being! (*Nulla est epigenesis*). No part of the animal body was made previous to another, and all were created simultaneously." As Haeckel (1897, vol. 1, p. 39) observed, Haller went so far as to maintain even the existence of a beard in the newborn male, and the existence of horns in the hornless fawn; "all the parts were already present in a complete state, but hidden for a while from the human eye." Given Haller's enormous scientific stature in the 1700s, we can only assume that he had an overriding set about the question of ontogenesis (development of the individual) and that set caused him to misinterpret evidence in a selective way. For example, the strongest evidence for the theory of encasement, as the theory of preformation was sometimes called, derived from Bonnet's observations, in 1745, of virgin plant lice, who, without the benefit of a male consort, reproduce parthenogenetically (i.e., by means of self-fertilization). Thus, one can imagine the ovist Bonnet's excitement upon observing a virgin female plant louse give birth to ninety-five females in a twenty-one-day period and, even more strikingly, observing these offspring themselves reproduce without male contact. Here was Eve incarnate among the plant lice!

Epigenesis: Emergent Nature of Individual Development

But the empirical solution of the preformation–epigenesis controversy necessitated direct observation of the course of individual development, and not the outcome of parthenogenetic reproduction, as striking as that fact itself might be. Thus, it was that one Caspar Friedrich Wolff (1733–1794), having examined the developmental anatomy and physiology of chick embryos at various times after incubation, provided the necessary direct evidence for the epigenetic or emergent aspect of individual development. According to Wolff's observations, the different organic systems of the embryo are formed and completed successively; first, the nervous system, then the skin covering of the embryo, third the vascular system, and finally the intestinal canal. These observations not only eventually toppled the doctrine of preformation but also provided the basis for the foundation of the science of embryology, which took off in a very important way in the next 150 years. Those who like to dwell on the inscrutable coincidences of temporal cycles will undoubtedly be taken by several noteworthy events that followed at regular fifty-year intervals the publication of Wolff's doctoral dissertation in 1759; namely, in 1809 the birth of Charles Darwin and, in the same year, the publication of J. B. Lamarck's *Philosophical Zoology,* followed precisely fifty years later, in 1859, by the publication of Charles Darwin's own *Origin of Species.*

Before turning to the views of Lamarck and Darwin on evolution, it will be helpful to describe the fruits that resulted from the scientific study of embryology in the early 1800s, especially the work of Karl Ernst von Baer.

Karl Ernst von Baer: The Science of Embryology

Karl Ernst von Baer (1792–1876) directly examined the physiological and anatomical development of embryos and fetuses in many different species of mammals, birds, fishes, and invertebrates, and examined them more intensively than possibly any other scientist before or since his time. He published his observations—and his reflections on his observations—in two volumes of his big book, *Ueber Entwickelungsgeschichte der Thiere,* the first volume being published in 1828 and the second in 1837. Von Baer's systematic observations, and the laws of individual development that he correctly generalized from them, had an overwhelming significance in that they gave the necessary impetus to the beginning science of embryology and put an end to the preformation–epigenesis debate. Von Baer's work also had a convoluted influence on Charles Darwin's formulation of his theory of evolution and Ernst Haeckel's biogenetic law that said (incorrectly) that ontogeny recapitulates phylogeny.

Von Baer's observations signaled the end of the epigenesis–preformation debate because he observed over and over again that all of the various parts and functions of the embryo and fetus arise successively (i.e., epigenetically) during the course of embryonic and fetal development in all of the various species he observed. Von Baer is very specific on this point, stating that "All is transformation, nothing is development *de novo*" (translated by Oppenheimer, 1967, from von Baer, 1828, p. 156). For example, von Baer said that when nerves are formed there was not an empty space in the tissue beforehand but that nerves differentiated out of what was formerly homogeneous tissue. He began his observations at the earliest stages of development when the organism is composed of the initial germ layers, from which arise all of the various distinctive components of the adult animal. One of von Baer's colleagues, Christian Pander, had discovered the three germinal layers of the chick embryo in 1817. Von Baer extended the validity of Pander's discovery to other vertebrate embryos and, as Jane Oppenheimer (1940) has pointed out, thus provided the basis of comparative embryology from von Baer's time until today. From the standpoint of the preformation–epigenesis debate, the four "laws" that von Baer generalized from his empirical observations are of utmost importance.

Fig. 1–1. Karl Ernst von Baer (1792–1876).

Ontogeny Recapitulates Ontogeny (Up to a Point)

The title of the section of von Baer's book (1828, volume 1, p. 224) containing the four laws reads rather matter-of-factly: "The individual development of higher animal forms does not pass through the fully developed [i.e., adult] forms of lower animals." Since the German evolutionist Ernst Haeckel was a great admirer of von Baer's research, and no doubt had full command of the German language in which von Baer promulgated his laws, it seems incredible that Haeckel could have formulated his infamous biogenetic law that ontogeny recapitulates phylogeny, given von Baer's introductory comment and his laws of embryonic development. At any rate, here follows a free translation of von Baer's laws, which not only describe the epigenetic character of individual embryonic development but also describe the comparative relationship among embryos of related species:

> 1. The features that larger animal groups have in common appear earlier in embryos than do the more special features. [As an example, von Baer said the avian embryo first looks like a vertebrate before it looks like a bird, then it looks like a gallinaceous bird, and, finally, a specific kind of bird, e.g., a chicken.]
> 2. From the most general [features] of the form patterns arise the less general, and so forth, until finally the most specialized [features] appear [in ontogeny].
> 3. Each embryo of a particular animal species, instead of passing through the [adult] forms of certain other animals, rather differentiates itself from them.
> 4. Thus, basically the embryo of a higher animal species is never like [the adult] of another species but only like its embryo.

The theme that pervades von Baer's volumes is that individual ontogenetic development proceeds from the general to the special, or from initial homogeneity in the early embryo to heterogeneity in the fully developed adult. "All is transformation, nothing is development *de novo*." Thus, von Baer saw individual development as involving *differentiation,* as necessitated by the epigenetic theory, and not merely growth, as called for in the preformation doctrine.

So, by the time of the nineteenth century, the concept of epigenesis triumphed, as it were, over the concept of preformation. However, as we shall see in later chapters, wrongheaded thinking about the process of epigenesis itself kept alive a preformation-like predeterministic view of individual development that has persisted to the present day, and which is at the heart of the false nature–nurture dichotomy. As I will try to show in succeeding chapters, an overly simplistic and erroneous predeterministic view of epigenesis—one that holds that traits are caused by genes in a straightforward unidirectional manner—has been encouraged by intellectual developments in various quarters during the nineteenth and twentieth centuries. Among those to be described are Ernst Haeckel's and Charles Darwin's recapitulation doctrine; August

Weismann's separation of germ cells and body cells; Galton's reification of nature and nurture as two separate entities; the rediscovery of Mendel's ratios of hereditary transmission and the subsequent burgeoning of the field of population genetics, which does not take into account the complexities of individual development. Along the way, experimental embryology—the analytical study of individual development—became separated from the study of genetics, so real had the separate existence of nature and nurture become. It is the purpose of the present volume to recount these intellectual developments with a view to putting Humpty-Dumpty back together again.

Since behavioral and psychological development are at the heart of the present synthesis, it is appropriate to pay intellectual homage to the much maligned French thinker Jean Baptiste Lamarck in the chapter to follow.

◇ ◇ 2 ◇ ◇

Lamarck and the Idea of the Evolution of Species

Jean Baptiste Lamarck (1744–1829) was the first major proponent of the idea of evolution at the species level. His most important work for the present purpose was his treatise on *Zoological Philosophy*, which was first published in France in 1809 and translated into English in 1914 (reissued in 1984). It was in that book that Lamarck laid out his ideas and his arguments for the highly controversial notion that species are mutable. That is, Lamarck held that species can evolve and are not permanently fixed, as most naturalists of the time believed they were. It is a terrible irony that Lamarck's name is irrevocably connected with the concept of the inheritance of acquired characteristics, whereas his truly original conception was the idea of organic evolution at the species level. For Lamarck the concept of evolution was the way to understand the eventual emergence of higher psychological functions such as attention, thinking, memory, judgment, imagination, and reasoning. As Richard W. Burkhardt, Jr. (1977), one of Lamarck's biographers, has made clear, Lamarck was above all a systematic thinker ever on the alert for the large facts: "the great truths, which the philosophers could not discover because they had not sufficiently observed nature, and which the zoologists have not perceived because they

Fig. 2–1. Jean Baptiste Lamarck (1744–1829). Painted near the end of his life, after he had become blind.

have occupied themselves too much with matters of detail" (Lamarck, 1809/1984).

Behavioral Adaptability

For Lamarck behavioral adaptability plays a central role in the evolutionary changes in bodily form—it is habits that bring about morphological changes and not the other way around. No writer before or since has made behavior play such an explicitly important role in evolution. Similarly, no writer before or since has made behavioral adaptability such a central concept in fostering evolutionary change. It was Lamarck's view that changes in the environment create challenges and opportunities that are met by animals changing their behavioral habits. The changes in habit eventually change the bodies of the animals. Here is the way Lamarck works up to the formulation of his two laws of nature that bring about evolutionary change (Lamarck, 1809/1984, pp. 112–113):

> It is obvious then that as regards the character and situation of the substances which occupy the various parts of the earth's surface, there exists a variety of environmental factors which induces a corresponding variety in the shapes and structure of animals, independent of that special variety which necessarily results from the progress of the complexity of organism in each animal . . .
>
> Now the true principle to be noted in all this is as follows:
>
> 1. Every fairly considerable and permanent alteration in the environment of any race of animals works a real alteration in the needs of that race.
>
> 2. Every change in the needs of animals necessitates new activities on their part for the satisfaction of those needs, and hence new habits.
>
> 3. Every new need, necessitating new activities for its satisfaction, requires the animal, either to make more frequent use of some of its parts which it previously used less, and thus greatly to develop and enlarge them; or else to make use of entirely new parts, to which the needs have imperceptibly given birth by efforts of its inner feeling; this I shall shortly prove by means of known facts.
>
> Thus to obtain a knowledge of the true causes of that great diversity of shapes and habits found in the various known animals, we must reflect that the infinitely diversified but slowly changing environment in which the animals of each race have successively been placed, has involved each of them in new needs and corresponding alterations in their habits. This is a truth which, once recognised, cannot be disputed. Now we shall easily discern how the new needs may have been satisfied, and the new habits acquired, if we pay attention to the two following laws of nature, which are always verified by observation.
>
> First Law.
>
> *In every animal which has not passed the limit of its development, a more frequent and continuous use of any organ gradually*

strengthens, develops and enlarges that organ, and gives it a power proportional to the length of time it has been so used; while the permanent disuse of any organ imperceptibly weakens and deteriorates it, and progressively diminishes its functional capacity, until it finally disappears.

Second Law.

All the acquisitions or losses wrought by nature on individuals, through the influence of the environment in which their race has long been placed, and hence through the influence of the predominant use or permanent disuse of any organ; all these are preserved by reproduction to the new individuals which arise, provided that the acquired modifications are common to both sexes, or at least to the individuals which produce the young.[1]

It was very important to Lamarck to be able to come up with what he considered evidence for his assertions that the continued disuse and use of organs through changes of habit lead eventually to their decline or disappearance and to their functional perfection, respectively. He saw the absence of teeth in certain vertebrate species, for example, as a consequence of the species having long been exposed to an environment that induced the habit of swallowing food objects without any preliminary mastication. Under these conditions, according to Lamarck, the teeth regress and remain hidden in the bony framework of the jaws, or "they actually become extinct down to their last rudiments." Thus, Lamarck must have been delighted when his colleague Geoffroy Saint-Hilaire discovered teeth concealed in the jaws of the fetus of the right whale, a species otherwise known to be devoid of teeth. The same Geoffroy Saint-Hilaire discovered in birds the dental groove in which teeth would ordinarily appear if birds had teeth. Lamarck's law of disuse was likewise invoked to explain the disappearance (or near disappearance) of eyes in burrowing animals such as moles or cave-dwelling fish "whose habits require a very small use of sight." According to Lamarck's reasoning, light does not penetrate everywhere, thus providing environments in which animals may eventually become sightless or eyeless because of the diminished need for, and exercise of, vision. Sound, on the other hand, does penetrate everywhere, so Lamarck predicted there would never be found an animal species of "the vertebrate Plan" that would not have an organ of hearing—the environment would never offer the opportunity not to utilize the sense of hearing.

While Lamarck believed that "the disuse of any organ modifies, reduces and finally extinguishes it, the constant use of any organ, accompanied by efforts to get the most out of it, strengthens and enlarges that organ, or creates new ones to carry on functions that have become necessary." According to this view, waterfowl have webbed feet because their habit of swimming after prey eventually caused the skin to join the digits of their feet. The same reasoning would apply to the webbed feet of frogs, sea turtles, otters, beavers, and so on. Lamarck pressed on with this manner of thinking in a way that his opponents who did not believe in the mutability of species would find ridiculous. For example, he evolved wading birds with long legs and long necks by imputing to them an aversion to getting their bodies wet in the water in which they must fish for their prey.

> Now this bird tries to act in such a way that its body should not be immersed in the liquid, and hence makes its best efforts to stretch and lengthen its legs. The long-established habit acquired by this bird and all its race . . . results in the individuals . . . becoming raised as though on stilts, and gradually obtaining long, bare legs, denuded of feathers up to the thighs and often higher still. We note again that this same bird wants to fish without wetting its body, and is thus obliged to make continued efforts to lengthen its neck. Now these habitual efforts in this individual and its race must have resulted in course of time in a remarkable lengthening, as indeed we actually find in the long necks of all water-side birds. (Lamarck, 1809/1984, pp. 119–120)

Although the ideas concerning the consequences of the use and disuse of organs and the inheritance of acquired characteristics had been around since at least the time of Aristotle (Zirkle, 1946), these concepts had never before been applied to the question of evolution—that was Lamarck's entirely original theoretical contribution (Burkhardt, 1977; Zirkle, 1946). Needless to say, those who believed species to be immutable did not find Lamarck's reasoning or his examples convincing—indeed, one had to be sympathetic to the idea of evolution in order to find Lamarck's thinking at all palatable. One who found Lamarck's evolutionary theory entirely unpalatable was Georges Cuvier, who was considered to be the greatest and most influential comparative anatomist of the period. Some three years after Lamarck's death, Cuvier's most sardonic of eulogies was read before the French Academy of Science on November 26, 1832. Cuvier introduces his memorable critique of Lamarck's scientific career by saying, "in sketching the life of one of our most celebrated naturalists, we have conceived it to be our duty, while bestowing the commendation they deserve on the great and useful works which science owes to him, likewise to give prominence to such of his productions in which too great indulgence of a lively imagination has led to results of a more questionable kind" (Cuvier in Lamarck, 1809/1984, p. 435). Inevitably, Cuvier turns his attention to Lamarck's two evolutionary laws and comments on them rather unfairly, having conveniently overlooked Lamarck's caveat concerning the outcome of evolution being strictly constrained by the limits of what we today would call the potential or range of the species' developmental plasticity. Cuvier writes:

> These principles once admitted, it will easily be perceived that nothing is wanting but time and circumstances to enable a monad or a polypus gradually and indifferently to transform themselves into a frog, a stork, or an elephant. But it will also be perceived that M. de Lamarck could not fail to come to the conclusion that species do not exist [unchangingly] in nature; and he likewise affirms, that if mankind think otherwise, they have been led to do so only from the length of time which has been necessary to bring about those innumerable varieties of form in which living nature now appears. This result ought to have been a very painful one to a naturalist, nearly the whole of whose long life had been devoted to the determination of what had hitherto been believed to be species . . . and whose acknowledged merit, it must be confessed, consisted in this very determination. . . . A system established on such foundations may amuse the imagination of a poet; a metaphysician may derive from it an entirely new series of systems; but it cannot for a moment bear the examination of any one who has dissected a hand, a viscus, or even a feather. (Cuvier in Lamarck, 1809/1984, p. 446–447)

Cuvier writes here as if the course of evolution or nonevolution can somehow be ascertained merely by dissection of animals and their parts. The course of evolution (or nonevolution), then as now, can only be rendered by shrewd and educated inference weighing considerably more evidence than anatomy as such. Cuvier seems to be saying that firsthand acquaintance with anatomy would convince one that evolution is impossible, as well it might!

As we shall see in the ensuing chapters, even though his theoretical views were severely criticized by the great Cuvier, Lamarck's ideas survived, especially in the writings of Charles Darwin, his great successor in establishing the theory of evolution via natural selection *and* the inheritance of acquired characters. Among the specific ideas of Lamarck that has endured during the 150 years following his death was his belief that evolution is by and large directional and progressive, proceeding from the simple to the complex in bodily, behavioral, and psychological organization. Even though the fossil record (to which Lamarck had access) was largely incomplete, in broad outline it would testify to increasing skeletal complexity throughout the course of evolution from invertebrate to vertebrate animals, even if such progressive changes are not always and everywhere linear. Lamarck's emphasis on the adaptability of organisms to changing environments, plus the presumed ability to transmit improvements from one generation to the next in a cumulative way, meshes admirably with the observed general evolutionary increase in organismic complexity. This must have been a most satisfying intellectual solution to Jean Baptiste Lamarck, an inveterate systematic thinker and grand synthesizer of nature. It is fitting to conclude with a quotation from Lamarck himself (1809/1984, p. 126):

> Nature has produced all the species of animals in succession, beginning with the most imperfect or simplest, and ending with the most perfect, so as to create a gradually increasing complexity in their organization; these animals have spread at large throughout all the habitable regions of the globe, and every species has derived from its environment those habits that we find in it and the structural modifications which observation shows us.

Ernst Haeckel, among many others, recognized Lamarck's priority in clearly stating the theory of descent or transmutation, as evolution was called in the nineteenth century. Here is the way Haeckel stated it in 1897 (vol. 1, pp. 84–85):

> All the various species of animals and plants which we now see around us, or whichever existed, have developed in a natural manner from previously existing, different species; all are descendants of a single ancestral form, or at least of a few common forms. The most

ancient ancestral forms must have been very simple organisms of the lowest grade, and must have originated from inorganic matter by means of spontaneous generation. Adaptation through practice and habit, to the changing external conditions of life, has ever been the cause of changes in the nature of organic species, and heredity caused the transmission of these modifications to their descendants.

These are the principal outlines of the theory of Lamarck, now called the Theory of Descent or Transmutation, and to which, fifty years later, attention was again called by Darwin, who firmly supported it with new proofs. Lamarck, therefore, is the real founder of this Theory of Descent or Transmutation, and it is a mistake to attribute its origin to Darwin. Lamarck was the first to formulate the scientific theory of the natural origin of all organisms, including man, and at the same time to draw the two ultimate inferences from this theory: firstly, the doctrine of the origin of the most ancient organisms through spontaneous generation; and secondly, the descent of man from the mammal most clearly resembling man—the Ape.

What Darwin was to add to this theory was the very important principle of natural selection in the struggle for existence. In the end it was Darwin's principle of natural selection that was retained and the law of inheritance of acquired characters that was dropped, because it was believed that no mechanism of heredity existed by which the genetic basis for the acquired characters could be transmitted to succeeding generations. A more or less acceptable variation of that theme has survived, however; it is called *genetic assimilation*. Experiments by Conrad Waddington and others have demonstrated that selection for the ability to "acquire" a trait in the face of an environmental challenge eventually leads to the appearance of the trait in the absence of the environmental challenge, a phenomenon called the genetic assimilation of an acquired character. I shall describe this phenomenon more fully in Chapter 11, dealing with present-day views of the contribution of individual development to evolution.

To return to the theme above, getting Lamarck his due as the originator of the concept of evolution, and getting his name divorced from the notion that acquired characters are heritable, is a most difficult undertaking, given that very few people have a firsthand acquaintance with the literature and thinking in Lamarck's day. Here is the rather charming way Conway Zirkle expressed the problem of getting Lamarck appropriately recognized for his origination of evolutionary thinking:

> What Lamarck did was to accept the hypothesis that acquired characters were heritable, a notion which had been held almost universally for well over two thousand years and which his contemporaries accepted as a matter of course, and to assume that the results of such inheritance were cumulative from generation to generation, thus producing, in time, new species. His individual contribution to biological theory consisted in his application to the problem of the origin of species the view that acquired characters were inherited and in showing that evolution could be inferred logically from the accepted biological hypotheses. He would doubtless have been greatly astonished to learn that a belief in the inheritance of acquired characters is now labelled "Lamarckian," although he would almost certainly have felt flattered if evolution itself had been so designated. (Zirkle, 1946, pp. 91–92)

Finally, it is not that Lamarck was the first person to opt for evolution, nor even the first person to opt for evolution through the transmission of acquired modifications—Erasmus Darwin, Charles's grandfather, had already done that in 1794 in volume 1 of his *Zoonomia.* Lamarck just did it much better by dint of his acquaintance of many interrelated facts. As we shall see in Chapter 3, Charles Darwin did an even better job, using the additional hypothesis of natural selection and an even broader array of interrelated facts. What is also very important for the present purposes is Darwin's wrongheadedness about the role of individual development in evolution. He actually foreshadowed Haeckel's flawed notion that ontogeny recapitulates phylogeny, a view that remains with us in disguised form to the present day.

◇ ◇ 3 ◇ ◇

Charles Darwin on the Evolution of Species and the Role of Embryological Development

Charles Darwin was born in 1809, in the year that Lamarck's *Zoological Philosophy* was first published. He lived until 1882, managing to see his most famous book, *On the Origin of Species by Means of Natural Selection or the Preservation of Favoured Races in the Struggle for Life,* pass through six editions. The *Origin* was first published in 1859, but Darwin had been working sporadically on the manuscript since the early 1840s, soon after completing his five-year journey as naturalist aboard her majesty's ship the *Beagle* on its voyage around the world.

What impressed Darwin above all on this long voyage was the way species fitted their various environments or habitats so well through adaptations of their external organs and their behavior. His observations on the *Beagle* voyage eventually led him to believe that species gradually became modified (i.e., evolve), because each species that he saw, and its close relatives, was so exquisitely adapted to its surroundings. As he pondered the origin of these adaptations and the probability that species are mutable, he happened to read an essay by the economist Malthus who was worrying about the inevitable overpopulation of the world and the necessarily fierce competition that would ensue among human groups for the finite resources (food, living space, shelter, etc.) required to sustain life. Who would compete best would win (survive). Given that animals daily face the same problem—a struggle for life and existence—and given that some animals are better endowed than others to compete in the struggle for existence, Darwin saw natural circumstances favoring some individuals over others, and some species over others, thus leading to two eventualities: the gradual perfection of adaptations, on the one hand, and the extinction of unfit individuals and species, on the other. This was in essence the doctrine of the survival of the fittest, whether the fittest was an individual or a species. Thus, an individual or species that could change would do so as its conditions necessitated such change, or it would become extinct if it lacked the necessary variations to change in an adaptive way. The components of Darwin's theory of evolution are thus (1) individual variation, (2) competition, (3) selection, (4) increased adaptation or elimination.[1]

Early in the *Origin of Species,* here is the way Darwin expresses his great principle of natural selection and its relationship to evolution, though he does not explicitly use the word evolution here:

> Again, it may be asked how is it that varieties, which I have called incipient species, become ultimately converted into good and distinct species, which in most cases obviously differ from each other far more than do the varieties of the same species? How do those groups of species, which constitute what are called distinct genera, and which differ from each other more than do the species of the same genus, arise? All these results . . . follow inevitably from the struggle for life. Owing to this struggle for life, any variation, however slight and from whatever cause proceeding, if it be in any degree profitable to an individual of any species, in its infinitely complex relations to other organic beings and to external nature, will tend to the preservation of that individual, and will generally be inherited by its offspring. The offspring, also, will thus have a better chance of surviving, for, of the many individuals of any species which are periodically born, but a small number can survive. I have called this principle, by which each slight variation, if useful, is preserved, by the term of Natural

Fig. 3–1. Charles Darwin (1809–1882).

Selection to mark its relation to man's power of selection. [Darwin is here referring to the enormous proliferation of breeds of pigeons, dogs, etc., by man's exercise of artificial selection under conditions of domestication]. We have seen that man by selection can certainly produce great results, and can adapt organic beings to his own uses, through the accumulation of slight but useful variations, given to him by the hand of Nature. (C. Darwin, 1859, p. 61)

The essence of Darwin's concept of evolution is that it is slow and gradual—it does not go by leaps and bounds, and thus it does not leave big gaps. Consequently, his concept of evolution predicts

Fig. 3–2. Fossil remains of *Archaeopteryx,* showing reptilian and avian characteristics. (From de Beer, 1954.)

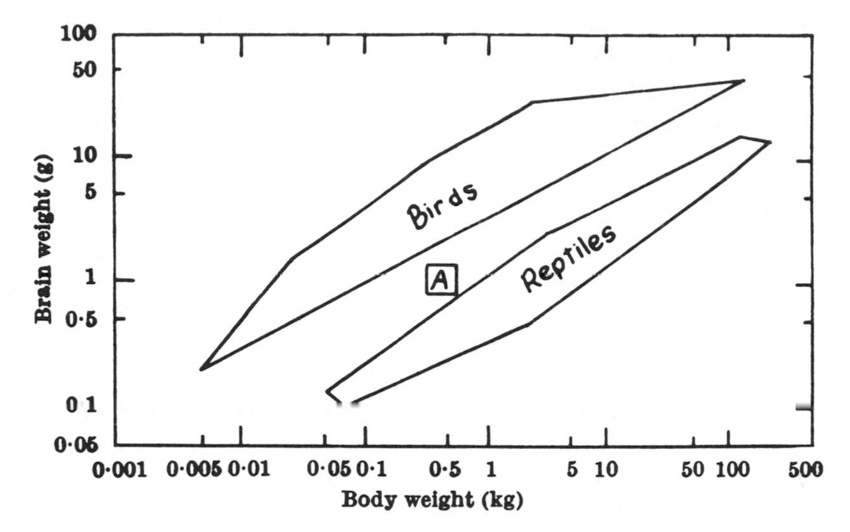

Fig. 3–3. The relative brain size of *Archaeopteryx* (A) compared to reptiles and birds of various brain and body weights. Its transitional status is well exemplified by its placement between reptiles (from which it evolved) and birds (the direction in which it was evolving). (From Jerison, 1968.)

the appearance of transitional forms linking ancestral taxonomic groups (say, reptiles) with succeeding taxonomic groups (say, birds). In this context one can imagine the excitement among supporters of the Darwinian theory of evolution when, in 1861, two years after the publication of the *Origin*, a fossil bird (*Archaeopteryx*) was uncovered in Germany that had a number of reptilian features (see Figure 3–2). For example, although this animal had wings, something like feathers, and hollow bones (avian characters), at the same time it possessed teeth and a relatively long bony tail (reptilian characters). And, as can be seen in Figure 3–3, its brain size relative to its body was midway between reptiles and birds, making it possibly the most splendid of transitional forms discovered to date (Jerison, 1968).[2]

The importance of finds such as *Archaeopteryx* cannot be overestimated because, as we now appreciate, Darwin's great book is nothing but a long, skillfully reasoned argument from beginning to end: evolution is so slow its process cannot be observed—we see only its end products, whether they are fossils or living specimens. One of the reasons for the persuasiveness of Darwin's theory of evolution via natural selection is that it accords well with a variety of facts or observable phenomena, and Darwin made the most of these concordances in writing and revising the text of his big book over a nearly twenty-year period. In the end it was a much more persuasive case than Lamarck was able to make for evolution—both the style and content of the arguments are quite different, even though Darwin did retain Lamarck's notions of use and disuse, and the inheritance of acquired characters, as contributing to the variation that is the stuff of evolution. In a very general way, Darwin also retained Lamarck's ideas of evolution as proceeding from one or few species to many, from the simple to the complex in organization, and from the less perfect to the more perfect, or from the lower to the higher, in form, function, behavior, and psychology. The latter themes in particular became the centerpiece of Herbert Spencer's theorizing about evolution not only in biology but in psychology and sociology (Richards, 1987).

Ontogeny Recapitulates Phylogeny

As I mentioned earlier, Darwin takes great pains to make his argument for evolution convincing and in this he spares no possible ally. Given our specific interests in the relationship between the concept of evolution and the concept of development, Darwin's treatment of embryology and embryological thinking is a case in point. Darwin's point of view is clearly depicted in the following quotation and, as we read it reflectively, we can understand his consternation, expressed in his *Autobiography* (Barlow, 1958, p. 125), that Ernst Haeckel should have received the credit for the biogenetic law, the view that ontogeny recapitulates phylogeny. This was written by Darwin in 1859 some seven years before Haeckel published his influential *Generelle Morphologie* in 1866.

> As all the organic beings, extinct and recent, which have ever lived on this earth have to be classed together, and as all have been connected by the finest gradations, the best, or indeed, . . . the only possible arrangement would be genealogical. Descent [evolution] being in my view the hidden bond of connexion which naturalists have been seeking under the term of the natural system [of the classification of animals]. On this view we can understand how it is that, in the eyes of

> most naturalists, the structure of the embryo is even more important for classification than that of the adult. *For the embryo is the animal in its less modified state; and in so far it reveals the structure of its progenitor.* In two groups of animal, however much they may at present differ from each other in structure and habits, if they pass through the same or similar embryonic stages, we may feel assured that they have both descended from the same or nearly similar parents, and are therefore in that degree closely related. Thus, community in embryonic structure reveals community of descent. It will reveal this community of descent, however much the structure of the adult may have been modified and obscured . . . as the embryonic state of each species and group of species partially shows us the structure of their less modified ancient progenitors, we can clearly see why ancient and extinct forms of life should resemble the embryos of their descendants,—our existing species. (Darwin, 1859, pp. 448–449, italics added)

In the above quotation, if Darwin were to stick to the notion that community of descent is evidenced by resemblances among embryos, he would be standing on the modern conception that Karl Ernst von Baer's comparative embryological research established so well, as described in the first chapter. Instead, as indicated by the italicized sentence in the quotation, Darwin wants embryology to provide the record of evolutionary descent—that is, he wants the embryonic stages of existing species to depict the genealogical history of the species, thereby providing the kind of graphic and immediate evidence that the theory of evolution might have seemed to require in 1859. As we shall see in the next chapter when we take up the contribution of Ernst Haeckel, this recapitulationist way of thinking is not only not valid when taken as an all-embracing generalization, it presents the gravest possible obstacle to understanding the relationship of ontogeny to phylogeny, or of individual development to evolution.

As Jane Oppenheimer (1967) has so well documented, Darwin got his wrongheaded idea about the testimony of embryology from the naturalist Louis Agassiz, and he could not be dissuaded from it even by the arguments of Thomas Huxley, one of his most ardent and dedicated supporters. In fact, possibly because of the perceived demands of his evolutionary theory, Darwin apparently became more convinced of the correctness of the recapitulation doctrine as the years went by. In the first edition of the *Origin* in 1859 Darwin wrote (p. 338):

> Agassiz insists that ancient animals resemble to a certain extent the embryos of recent animals of the same classes; or that the geological succession of extinct forms is in some degree parallel to the embryological development of recent forms. I must follow Pictet and Huxley in thinking that the truth of this doctrine is very far from proved. Yet I fully expect to see it hereafter confirmed, at least in regard to subordinate groups, which have branched off from each other within comparatively recent time. For this doctrine of Agassiz accords well with the theory of natural selection.

Oppenheimer (1967, p. 254) tracked this statement from the first to the fifth edition of the *Origin* in 1872, where Darwin had emended it to read as follows:

> Agassiz and several other highly competent judges insist that ancient animals resemble to a certain extent the embryos of recent animals belonging to the same classes; and that the geological succession of extinct forms is nearly parallel with the embryological development of existing forms. This view accords admirably well with our theory.

As Oppeneheimer points out, Darwin did not have any of von Baer's works in his library at the time of his death in 1882. We also know from various sources that Darwin did not read German easily or with great understanding, thus von Baer's writings would have been quite a chore in the original. But German was Ernst Haeckel's mother tongue and still he managed to stand von Baer on his head, as we shall see in the next chapter.

◇ ◇ *4* ◇ ◇

Ernst Haeckel and the Biogenetic Law

Ernst Haeckel (1834–1919) was arguably one of the most influential biologists of the late 1800s. He was not only a consummate professional, he was also a popularizer: his very clearly written books were translated into twenty or more languages, and sold in the hundreds of thousands. Haeckel's most original scientific contribution was to develop to its fullest the idea expressed by Louis Agassiz and Charles Darwin: that embryological development parallels evolutionary development or that ontogeny recapitulates phylogeny. Here is the way Haeckel expressed his biogenetic law, which he considered to be the fundamental law of organic evolution:

> The History of the Germ is an epitome of the History of the Descent, or, in other words: that Ontogeny is a recapitulation of Phylogeny; or, somewhat more explicitly: that the series of forms through which the Individual Organism passes during its progress from the egg cell to its fully developed state, is a brief, compressed reproduction of the long series of forms through which the animal ancestors of that organism (or the ancestral forms of its species) have passed from the earliest periods of so-called organic creation down to the present time.
>
> The causal nature of the relation which connects the History of the Germ (Embryology, or Ontogeny) with that of the tribe (Phylogeny) is dependent on the phenomena of Heredity and Adaptation.

> When these are properly understood, and their fundamental importance in determining the forms of organisms recognized, we may go a step further, and say: Phylogenesis is the mechanical cause of Ontogenesis. The Evolution of the Tribe, which is dependent on the laws of Heredity and Adaptation, effects all the events which take place in the course of the Evolution of the Germ or Embryo. (Haeckel, 1897, vol. 1, pp. 6–7)

Haeckel's conception of the causal relation between evolution and the development of the individual, while at first seeming to be a valuable insight that not only makes a great deal of sense but aids our understanding of both evolution and ontogeny, turns out upon critical analysis to be entirely wrongheaded. Not only is it wrongheaded, it continues to play an important if unrecognized role in certain current misguided conceptions about the role of genes in individual development. Since this is a strong charge, I will quote further from Haeckel's own writings to make certain he has his say as I launch my critique.

It will be immediately recognized that if phylogeny were indeed the mechanical cause of ontogeny, and if ontogeny did represent an abbreviated recap of phylogeny, observation of individual embryonic development would produce the complete ancestral record of the species. If true, it would thus be a very convenient guide to the course of evolution, to say the least. One problem that Haeckel himself acknowledged is that ontogeny never faithfully reproduces the ancestral record; there are considerable gaps in the ontogeny of any species if that ontogeny is to be read as a road map of evolution. These gaps are unexplained by Haeckel.

> Indeed, there is always a complete parallelism between the two series of evolution. [It is telling that Haeckel often uses the term evolution to refer to ontogeny, the development of the individual, as well as to phylogeny, the evolution of the species.] This is vitiated by the fact that in most cases many forms which formerly existed and actually lived in the phylogenetic series are now wanting, and have been lost from the ontogenetic series of evolution. (Haeckel, 1897, vol. 1, p. 8)

Second, and this point was not understood by Haeckel, there often appear in ontogeny strangers that should not be there if ontogeny produces a faithful phylogenetic record.

Fig. 4–1. Ernst Haeckel (1834–1919).

From the fact that the human egg is a simple cell, we may at once infer that there has been at a very remote time a unicellular ancestor of the human race resembling an Amoeba. Again, from the fact that the human embryo originally consists merely of two simple germ-layers, we may at once safely infer that a very ancient ancestral form is represented by the two-layered Gastraea. A later embryonic form of

> the human being points with equal certainty to a primitive worm-like ancestral form which is related to the sea-squirts or Ascidians of the present day. But the low animal forms which constitute the ancestral line between the unicellular amoeba and the gastraea, and further between the gastraea and the ascidian form, can only be approximately conjectured with the aid of comparative Anatomy and Ontogeny. (Haeckel, 1897, vol. 1, p. 9)

Not bad you say? A lot of gaps but still helpful? I would ask you to notice what a pivotal role the gastraea play in the above construction. Haeckel realized the gastraea were an essential link if ontogeny is to supply any sort of record, even if very patchy, of phylogeny, so much so that he made them up! Yes, the gastraea do not exist, and have never existed, anywhere but in Haeckel's writings. If you reread the above quotation with the realization that the gastraea are a fiction, the phylogenetic reconstruction from ontogeny is woefully deficient. In fact, one would question whether the concept is really tenable. And the gastraea, that is, the two-germ-layer stage of ontogeny, is not the only important stranger to appear in vertebrate ontogeny. Here are some other important strangers in human ontogeny.

> In man, as in all other higher vertebrates, the following incidents of germ-history must be regarded as palingenetic [true ancestral lineal] processes: the formation of the two primary germ-layers [our gastraean friends], the appearance of a simple notochord between the spinal tube and the intestinal tube, the transitory formation of gill-arches and gill-openings, of primitive kidneys, of the primitive brain bladder, the hermaphrodite rudiment of the sexual organ, etc. All these, and many other important phenomena have evidently been accurately handed down, by constant heredity, from the primeval ancestors of Mammals, and must, therefore, be referred directly to corresponding palaeontological evolutionary incidents in the history of the tribe. On the other hand [here's the real trouble], this is not the case with the following germinal incidents, which must be explained as kenogenetic [usually written cenogenetic in English, signifying falsely testifying] processes; the formation of the yolk sak [*sic*], of the allantois and placenta, of the amnion and chorion, and, generally, of the different egg-membranes and the corresponding system of blood-vessels; also the transitory separation of the primitive vertebrate plates and the side-plates, the secondary closing of the stomach wall

> and the intestinal wall, the formation of the navel, etc. All these, and many other phenomena are evidently not referable to corresponding conditions of an earlier independent, and fully developed parent form, but must be explained as solely due to adaptation to the peculiar conditions of egg-life or embryo-life (Haeckel, 1897, vol. 1, pp. 11–12)

What Haeckel thought he was seeing—what he wanted to see—and what he actually was seeing are two very different things. Haeckel was seeing the common ontogeny of various vertebrates species—the events that are typical of vertebrate or of mammalian embryonic development. That is, instead of seeing phylogeny Haeckel was seeing ontogeny! He was so tuned in to noticing all the ontogenetic exceptions to the paleontological record that embryology today gratefully employs all the terms that Haeckel invented to explain away the false testimonials. Not only *palingenetic* and *cenogenetic,* but *heterotropic* and, best of all, *heterochronic.* Heterotropic refers to a displacement in space or place of some presumed ancestral event, such as the human sexual organs deriving from the middle germ layer instead of the inner and outer germ layers as in lower forms. The heterochronic displacements in time are now recognized as a fundamental ontogenetic mechanism for evolutionary change, whereas Haeckel saw them as another version of false testimonial or "vitiation," as he put it, of the phylogenetic record. Heterochronic changes in timing refer principally to the acceleration or retardation of the appearance of an organ. In man, for example, Haeckel himself recognized the ontogenetic acceleration of the heart, the gill-openings (not truly fish gills), brain, eyes, etc. Decelerations in timing include the relatively slow maturation of the intestinal tract and the body cavity, and the delay in achieving sexual (reproductive) competency. We now see these changes as being an integral part of evolution itself rather than seeing them as cenogenetic in Haeckel's sense of vitiating the phylogenetic record or falsely testifying to "the connected series of diverse animal forms." To put it another way, Haeckel saw ontogeny as occurring only in one mode: the evolution of each new species would be tacked on to the end of the ontogeny of its immediate ancestor. While this terminal addition is indeed one way evolution can proceed, it is not only not the sole way it proceeds, it is probably one of the lesser used routes to the production of new species. As we shall see in later chapters, it is the various changes in the timing of developmental events, or heterochrony, that is possibly the most common ontogenetic mode for bringing about evolutionary change. In general, then, it is changes in ontogeny that lead to evolution, and not the other way around as Haeckel thought.

Because the recapitulation idea is so important and has in fact persisted in veiled form to the present day in the thinking of so many biologists and psychologists, I think it is helpful to further document the reality of the idea for Haeckel by further quoting from his own writings (1897, vol. 1, p. 232):

> By now again applying our first principle of Biogeny [ontogeny recapitulates phylogeny], we immediately obtain the following very important conclusion: *Man and all those other animals, which in the first stages of their individual evolution* [read individual development] *pass through a two-layered structural stage of a Gastrula-form, must have descended from a primeval, simple parent-form, the whole body of which consisted throughout life, as now in the case of the lowest Plant-animals, only of two different cell-strata or germ-layers.* To this most important primeval parent-form . . . we will now provisionally give the name of the Gastraea (i.e., primitive intestinal animal). (italics in original)

The provisional or hypothetical nature of these imaginary animals was soon forgotten—given the truth of the biogenetic law that assumed such a necessary and indubitable reality that Haeckel not only placed them in a nonhypothetical progression from amoeba to man (see Figure 4–2), he also produced a picture of a histological section through the fictitious gastraea (Figure 4–3) without acknowledging its hypothetical nature (Haeckel, 1891, p. 161)!

Because almost the whole of late nineteenth century biology was under the sway of Haeckel's biogenetic law, the fictitious gastraea somehow became a reality as the progenitor of all multicellular organisms, displacing real animals such as the coelenterates

---

Fig. 4–2. Haeckel's evolutionary classification of animal groups, beginning at the bottom with single-celled creatures and culminating in humans ("*Menschen*") at the top of the tree. The hypothetical gastraea ("*Gastraeden*") are shown at the first transition point above single-celled organisms (fifth from bottom in trunk of tree). (From Haeckel, 1891.)

Stammbaum des Menschen. Taf.XV.

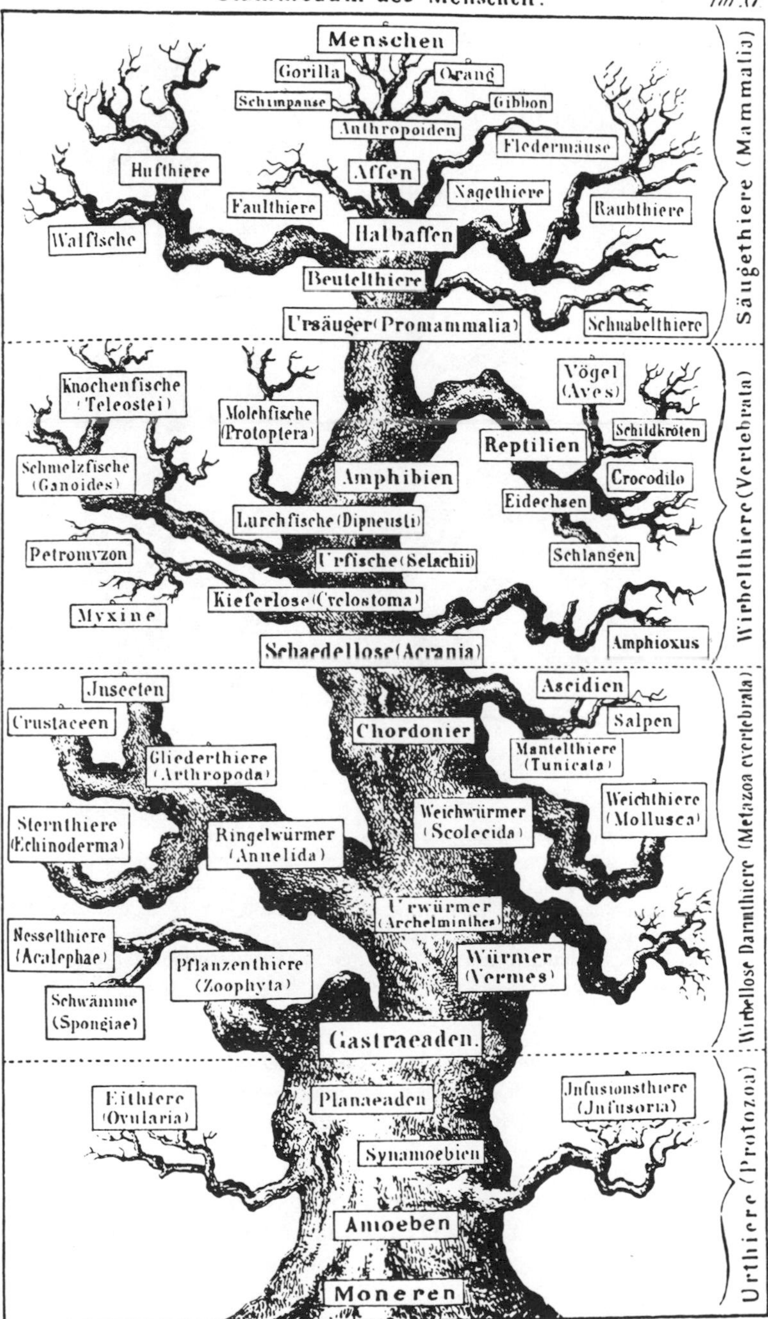

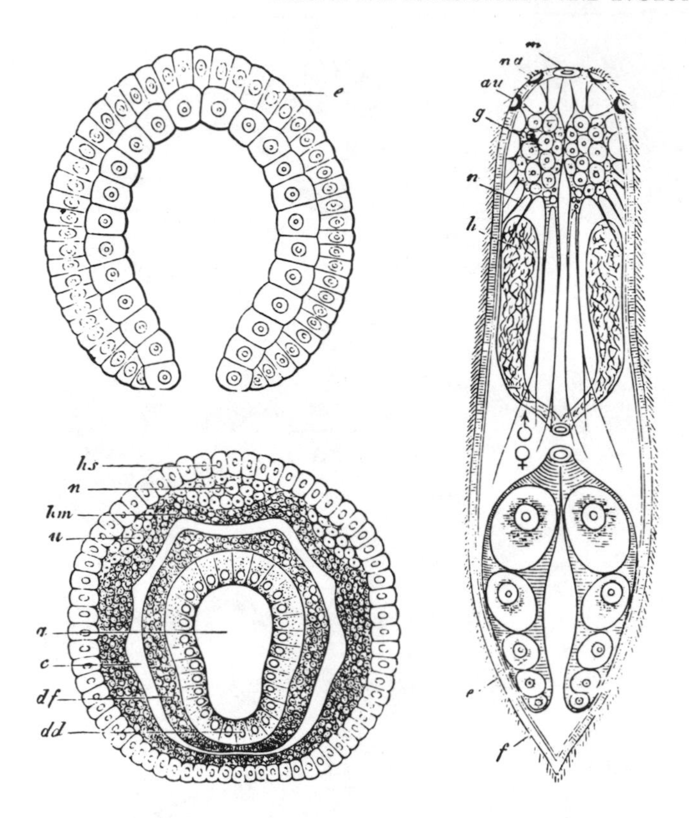

Fig. 209.—Gastrula of Gastrophysema (Gastræad-class).

Fig. 210.—Transverse section through an embryonic Earth-worm : *hs*, skin-sensory layer ; *hm*, skin-fibrous layer ; *df*, intestinal-fibrous layer ; *dd*, intestinal-glandular layer ; *a*, intestinal cavity ; *c*, body-cavity, or *Cœloma* ; *n*, nerve-ganglia ; *u*, primitive kidneys.

Fig. 211.—A Gliding Worm (*Rhabdocœlum*). From the brain or upper throat ganglion (*g*) nerves (*n*) radiate towards the skin (*f*), the eyes (*au*), the organ of smell (*na*), and the mouth (*m*) : *h*, testes ; *e*, ovaries.

Fig. 4–3. Hypothetical histological section of Haeckel's fictitious gastraead-class organisms (upper left), shown in company with histological drawings of real organisms (lower left and right). (From Haeckel, 1897.)

that had been previously nominated for that position! As Jane Oppenheimer (1967, pp. 150–151) puts it:

> It is one of the more curious ironies of history that while before Darwin, transcendentalism had closed the minds of investigators to

> the possibility of explanation of resemblances between parts of organisms and between whole organisms on the basis of common descent, yet after him the combination of the doctrines was to lead to extremes of exaggeration that were attained separately by neither. There were many who were to contribute to this: Kleinenberg in Germany and Lankester in England made an early start by relating phylogeny to ontogeny on the basis of comparability of the germ layers . . . , but their views were relatively mild compared to those of many who followed them. The culmination was the work of Haeckel, the greatest revisionist of them all.
>
> The most extreme example of his immoderation was perhaps his naming, on the basis of the "similarity, or *homology,* of the gastrula in all classes of compound animals" . . . the imaginary gastraea or the progenitor of all multicellular forms. His [drawing] of a section through an animal that never existed on the same page (1891, p. 161) that illustrates Kowalewski's gastrulae of Sagitta and Amphioxus and Carl Rabi's of Limnaeus, with no comment in the label to signify that the [gastraea] is any less real than the others—where is there a handsomer example in all biological or scientific history of what Whitehead has called the "Fallacy of Misplaced Concreteness"?
>
> Such a silly invention as the gastraea . . . as an isolated case might probably have proved of little influence; and its significance is as a symptom (a word used advisedly for its pathological connotations) of Haeckel's basic trouble. What was damaging to science was Haeckel's fervency to oversystematize all morphology through his biogenetic law that "die Ontogenie ist eine Recapitulation der Phylogenie."

In succeeding chapters we shall see how the recapitulation idea continues to rear its head to the present day, albeit in much revised form, in the thinking of some biologists and psychologists when they write about the genetic and evolutionary roots of behavior.

We cannot take leave of Ernst Haeckel without acknowledging his uncanny gift for coining terms that were found apt and acceptable by his fellow scientists. We owe to Haeckel not only the terms *ecology, phylum, ontogeny,* and *phylogeny,* but words to refer to the evolutionary or comparative aspect of embryological processes such as *heterochrony, cenogenesis,* and *palingenesis,* as well as the word for one of the universal early anatomical stages of embryonic development in multicellular animals, the *gastrula* stage, to name but a few of his coinages that immediately come to mind. The term that he would have most wanted (and, indeed, had every reason to expect) to be retained indefinitely was his *biogenetic law* to refer to the condensed recapitulation of phylogeny by ontogeny, but here he was to be denied, not the term, but of the generality of the evolutionary process with which his name shall always be linked.

Haeckel's Dubious Scientific Legacy

Since Haeckel believed phylogeny or evolution to be the mechanical cause of ontogeny, he was uncompromisingly hostile to the mechanical or causal analysis of ontogeny itself, which was the aim of the science of experimental embryology as led by Wilhelm His and Wilhelm Roux.

Haeckel ridiculed His's research from the start and he kept at it for twenty-five years or more. To understand correctly what was at stake it must be borne in mind that Haeckel believed there was only one true ontogenetic course of events, that which recapitulated phylogeny or the temporal evolutionary succession of animal forms. You'll recall his concept of cenogenesis to refer to ontogenetic displacements in time or space that vitiated the evolutionary testimony of any particular ontogeny that did not subscribe to strict recapitulation. Consequently, when His began mechanical analyses of the proximate or immediate causes of embryological development in the chick and traced out the very early development of the heart and vascular system, neither the sequence nor origin of the vascular tissues conformed to Haeckel's strict recapitulation of phylogeny: this was for Haeckel a false ontogeny because it did not testify to the succession of evolutionary events and it was therefore not only of no value, it was misleading. Haeckel refused to see that ontogenetic analysis could be saying something else—it had to testify to strict recapitulation to be valuable and that could be accomplished entirely by description, making the mechanical or causal analysis of ontogeny superfluous. Strict recapitulation means that organs have to originate from particular germ layers and have to make their temporal appearance in the embryo in parallel with their temporal appearance in phylogeny or evolution. As experimental enbryologists examined the early ontogenetic development of more and more different species, it gradually became apparent that the germ layers—ectoderm, mesoderm, entoderm—did not always give rise to the same organs in different species. Thus, it had to be recognized that evolution involved changes in ontogeny itself, a fact that

Haeckel could not accept because of the way he viewed the relationship of ontogeny to phylogeny: the latter was the mechanical cause of the former. As Garstang, de Beer, Goldschmidt, and others were to show in the early twentieth century, it is the mechanically caused changes in ontogeny that bring about evolution and not the other way around as Haeckel believed. This momentous change in the conception of the relationship of ontogeny and phylogeny will be discussed in later chapters. The full conceptual significance of this change in viewpoint—that changes in ontogeny are the basis for evolutionary change—is yet to be realized.

If we wish to be as appropriately generous in recognizing Haeckel's contribution to evolutionary thinking as we have in describing Lamarck's otherwise ridiculed views, it is to his lasting credit that he got the paleontologists—some of them at least—to go beyond mere classification and to begin to speculate about actual genealogical lines of phylogenetic descent, something that was not highly developed when Haeckel wrote his *Generelle Morphologie* in 1866.

◇ ◇ 5 ◇ ◇

St. George Mivart: First Intimations of the Role of Individual Development in Evolution

St. George Mivart (1827–1900) was an unlikely scholarly product, being an admirer of Richard Owen and a protégé of Thomas Huxley, archrivals in the paleontological field in nineteenth century England (Desmond, 1982). Most important, he appears to have been the most effective scientific critic of natural selection as the main mechanism of evolution, originating a line of argument that persists to the present day in the writings of Leo Berg, Richard Goldschmidt, Sewall Wright, Pere Alberch, Niles Eldridge, Stephen Gould, Soren Løvtrup, and others. As testimony to Mivart's effectiveness, he is the only one of Darwin's numerous critics to be singled out for mention in Darwin's *Autobiography* (p. 126). Darwin hides behind a comment from an American admirer to label Mivart a "pettifogger," so it is clear that he really did get under Charles's skin.

The title of Mivart's big book has a significant dual edge to it:

Fig. 5–1. St. George Mivart (1827–1900).

On the Genesis of Species. It was published in 1871, just twelve years after the original publication of Darwin's *Origin*. The double edge to *Genesis* involves, on the one hand, an unstinting assault on the *Origin* as not dealing with the origination of any species and, second, the invoking of a more or less unfathomable mechanism whereby species arise suddenly with large-scale, harmonious changes already intact. This was in stark contrast to Darwin's gradual selection of slight favorable variations to produce a new species only after a very long period of such selection. Throughout his career, Mivart tried to harmonize evolution with a modified and updated version of Judeo–Christian theology. He was a religious person, who converted to Roman Catholicism in his teens, only to be excommunicated six weeks before he died, because he publicly resisted the newly introduced notion of papal infallibility (Gruber, 1960).

In the present context, Mivart is of particular interest because he appears to be the first systematic thinker to make individual ontogenetic development central to his view of the basis for evolutionary change. Like most writers who followed him in this vein, Mivart stressed the definite or constrained nature of individual variation, so that such variations were neither random nor indefinite, as they were in Darwin's version. He brought this point of view to undermine Darwin's (and Alfred Russel Wallace's) account of mimicry, in which an otherwise palatable species of prey comes to look and act like an unpalatable species and thereby favors its survival. In other cases of mimicry, organisms tend to take on camouflage so they look like plants or some other nondescript object, which enhances their freedom from predation. In considering the evolution of mimicry, Mivart (1871, pp. 33–34) says:

> Now let us suppose that the ancestors of these various animals were all destitute of the very special protections they at present possess, as on the Darwinian hypothesis we must do. Let it also be conceded that small deviations from the antecedent colouring or form would tend to make some of their ancestors escape destruction by causing them more or less frequently to be passed over, or mistaken by their persecutors. Yet the deviation must, as the event has shown, in each case be in some definite direction, whether it be towards some other animal or plant, or towards some dead or inorganic matter. But as, according to Mr. Darwin's theory, there is a constant tendency to indefinite variation, and as the minute incipient variations will be in

all directions, they must tend to neutralize each other, and at first to form such unstable modifications that it is difficult, if not impossible, to see how such indefinite oscillations of infinitesimal beginnings can ever build up a sufficiently appreciable resemblance to a leaf, bamboo, or other object, for "Natural Selection" to seize upon and perpetuate. This difficulty is augmented when we consider . . . how necessary it is that many individuals should be similarly modified simultaneously.

Mivart is not interested in the complete overthrow of natural selection but merely wishes to demote it to the subordinate context of operating along with many other known and unknown causes or influences, which he explicitly labels as "obscure and mysterious." He vies for an internal natural law that causes organisms to change rapidly and harmoniously: "the efficient presence of an unknown internal natural law or laws conditioning the evolution of new specific forms from preceding ones, modified by the action of surrounding conditions, by 'Natural Selection' and by other controlling influences" (p. 45).

So Mivart, like Thomas Huxley before him, believes in evolution as a fact without believing in natural selection as its main or most formidable agency. The difference between the men is that Mivart actually put forth an alternative theory of evolutionary change, whereas Huxley played the role of evolution's most polished and influential spokesman and defender (Desmond, 1982). Also, in agreement with Mivart's point of view, Huxley (1870) was able to say, in his *Lay Sermons,* that Nature "does make considerable jumps in the way of variation now and then, and that these saltations give rise to some of the gaps which appear to exist in the series of known forms."

As stated previously, Mivart differs from Darwin in his (Mivart's) stress on the internally constrained variation of each species. He also differs with Darwin in the way he sees changes in the environment operating. With his eye always on the individual, especially the environing conditions of individual development, Mivart sees severe changes in the environment as bringing out new phenotypes or variations as a consequence of an alteration in the conditions of development: this is the developing organism's direct response to altered conditions of the developmental medium. Thus, Mivart takes Darwin's enormous catalogue of changes in plants and animals under domestication as an occasion to point out that the enormous change in environing conditions brings out new variations on which selection can now operate—Mivart's point throughout his book being that natural selection works on "what was born fit." Mivart also takes pains to point out that such changes are not infinite and are strictly limited by unknown internal factors. He does see the observed changes as being what we would today call "species-typical," that is, taking characteristic form in each species under given environmental conditions. Plasticity or malleability itself is a species-specific character, so Mivart notes that some species are more constant or inflexible in their organization. Darwin himself had demonstrated the rather extreme variability of dogs, horses, fowl, and pigeons under domestication, but he also noted the small degree to which the goose, peacock, and guinea-fowl changed under similar conditions. That is, there are many more varieties of the former than of the latter. So, once again, Mivart takes Darwin's factual observations and makes very different theoretical points concerning the direct effects of the environment altering the phenotype, and the constrained, directional, and species-typical nature of the changes. Once again, selection is relegated to a minor role—other factors generate or originate the variations upon which selection can then work and the variations themselves are not fortuitous or indefinite.

A specific example of the difference in the viewpoints of Darwin and Mivart is indicated by Darwin's comment, "In regard to fish, I believe that the same species never occur in the fresh waters of distant continents" (cited in Mivart, p. 145). This would follow from Darwin's notion of the occurrence of more or less indefinite variations that would then become fixed (selected) in relation to the particular nature of local conditions. In Mivart's view freshwater fish would tend to vary in specific and definite ways, and thus one would be likely to find the same species arising on remote continents in somewhat different habitats or local conditions. So Mivart is obviously delighted when one Dr. Gunther discovers the same genus of freshwater Indian fish on the coast of Western Africa, and other common East Indies freshwater fish in the Upper Nile and again in West Africa. The best case: "The genus *Galaxias* has at least one species common to New Zealand and South America, and one common to South America and Tasmania. In this genus we thus have an absolutely and completely fresh-water form *of the very same species* distributed between different and distinct geographical regions" (Mivart, 1871, p. 147, italics in original). Darwin was not unaware of these and other similar findings, which were anomalous from his point of view, and he attempted to understand them as originations from primordial, once singular land masses or transfers of fauna from one place to another by glaciers. Thus, these do not represent insurmountable difficulties for Darwin's theory, as Mivart acknowledges: "We have seen, then, that the geographical distribution of animals presents difficulties, though not insuperable ones, for the Darwinian hypothesis. . . . These facts, however, are not opposed to the doctrine of evolution; and if it could be established that closely similar forms had really arisen in complete independence one of the other, they would rather tend to strengthen and to support that theory" (Mivart, pp. 153–154).

One of Mivart's strongest arguments for definite or directional constraints on development leading to similar structural arrangements in otherwise very different species is the fact of homology. Examples of a homology well known to everyone are the resemblance among the arm of humans, the foreleg of a horse, the front flipper of whales, and the wing of a bat (a mammal) and that of a bird. These structural arrangements, though differing in detail, all appear in the same relative position to the whole of the animal's body and to surrounding parts; they are likely derived from a common ancestor, and, most important for the present purposes, they are similar in their developmental mode of origin. Here is the way Mivart (p. 156) put it:

> Now, it is here contended that the relationships borne one to another by various component parts, imply the existence of some innate, internal condition, conveniently spoken of as a power or tendency, which is quite as mysterious as is any innate condition, power, or tendency, resulting in the orderly evolution of successive specific manifestations. These relationships, as also this developmental power, will doubtless, in a certain sense, be somewhat further explained as science advances. But the result will be merely a shifting of the inexplicability a point backwards, by the intercalation of another step between the action of the internal condition or power and its external result. . . . It is not improbable that, could we arrive at the causes conditioning all the complex inter-relations between the several parts of one animal, we should at the same time obtain the key to unlock the secrets of specific origination.

It is here that Herbert Spencer enters our story, because he was the first to point out that homologous structures result from the similarity in the influences and conditions to which they are exposed during the course of individual development. To this Mivart adds that there must also be some similarity of internal tendency to respond in a certain way to these influences and conditions:

> Some internal conditions (or in ordinary language some internal power and force) must be conceded to living organisms, otherwise incident forces must act upon them and upon nonliving aggregations of matter in the same way and with similar effects.
>
> If the mere presence of these incident forces produces so ready a response in animals and plants, it must be that there are, in their case, conditions disposing and enabling them so to respond . . . as the same rays of light which bleach a piece of silk, blacken nitrate of silver. If, therefore, we attribute the forms of organisms to the action of external conditions, *i.e.*, of incident forces on their modifiable structure, we give but a partial account of the matter, removing a step back, as it were, the action of the internal condition, power, or force which must be conceived as occasioning such ready modifiability. (Mivart, pp. 166–167)

On this sort of account, Mivart believed it was easier to understand the same eye structure arising independently in vertebrate animals and in invertebrate molluscs, or a "curious resemblance" between the auditory organs of fishes and the invertebrate cephalopods which, again, could not be due to "genetic affinity" (Mivart's words).[1] It was difficult to accept these resemblances arising by chance variations in such widely separated groups, as necessitated by Darwin's theory of natural selection.

Genesis of Change: Individual Development

As has been said repeatedly in this chapter, Mivart believed in a non-Darwinian basis of evolution, in which natural selection was only one of a number of secondary causes or agencies. According to Mivart, the evolution of new species occurs by changes in individual development of preexisting species, one class of these developmental changes being what we today would call mutations and the other class being due to hereditary peculiarities of the parents. We must remember, or course, that Mivart was writing thirty years before the rediscovery of Mendel's work on the laws of inheritance in garden peas, the naming of genes as the carriers of inheritance, and the known facts of mutation or mutability during individual development, all of which occurred around 1900. Here is the way Mivart expressed himself on the origin of species:

> Now the new forms must be produced by changes taking place in organisms in, after or before their birth, either in their embryonic, or towards or in their adult, condition.
>
> Examples of strange births are sufficiently common, and they may arise either from direct embryonic modifications or apparently from some obscure change in the parental action. To the former category belong the hosts of instances of malformation through arrest of development, and perhaps generally monstrosities of some sort are the result of such affections of the embryo. To the second category [parental action] belong all cases of hybridism, of cross breed, and in all probability the new varieties and forms. . . . In all these cases we do not have abortions or monstrosities, but more or less harmonious forms often of great functional activity, endowed with marked viability and generative prepotency, except in the case of hybrids, when we often find even a more marked generative impotency.
>
> It seems probable therefore that new species may arise from some constitutional affection of parental forms—an affection mainly, if not exclusively, of their generative system. Mr. Darwin has carefully collected numerous instances to show how excessively sensitive to various influences this system is. . . . The gradually accumulating or diversely combining actions of all these [influences] on highly sensitive structures, which are themselves possessed of internal responsive powers and tendencies, may well result in occasional repeated productions of forms harmonious and vigorous, and differing from the parental forms in proportion to the result of the combining or conflicting action of all external and internal influences.
>
> If, in the past history of this planet, more causes ever intervened, or intervened more energetically than at present, we might *a priori* expect a richer and more various evolution of forms more radically differing than any which could be produced under conditions of more perfect equilibrium. (Mivart, 1871, pp. 233–235)

In compliance with his belief that the gaps in the fossil record are a datum, Mivart held that new species come into existence suddenly and completely, "not minute and infinitesimal." (See Fig-

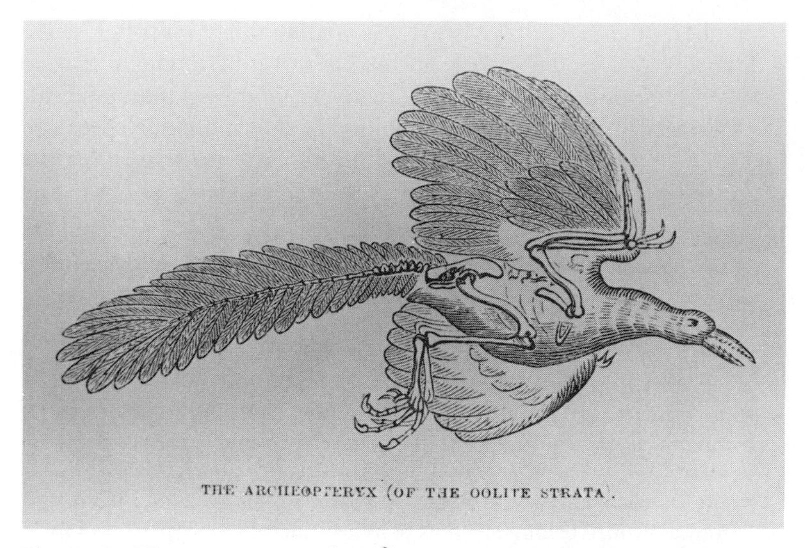

Fig. 5–2. The appearance of *Archaeopteryx* as reconstructed from the fossil record—half bird, half reptile—supported Mivart's notion that new species come into existence suddenly and completely. This animal had a bony reptilian tail with feathers on it, as well as feathers and claws on its wings. (From Mivart, 1871.)

ure 5–2.) Consistent with his earlier observations, Mivart believed that the capacity or tendency to change was itself heritable and thus was passed on in the process of evolution. In sum, Mivart believed that evolution was brought about by the united action of internal and external forces that serve to change individual ontogenetic development, sometimes resulting in abortions and monstrosities, and, at other times, in harmonious ("self-consistent") new organisms. Mivart contrasts his view to that of Richard Owen, who believed that "every species changes in time, by virtue of inherent tendencies thereto." Owen called his delimited concept of evolution "derivation" and viewed each species as having an "innate tendency to change irrespective of altered circumstances" (Desmond, 1982). Mivart expressly ascribed greater weight to external influences, operating in conjunction with internal ones, in bringing about evolution.

While the power of Darwin's concept of natural selection is that it specifies the variety and nature of external influences that guide and determine evolution, it says nothing about how the changes originate. ("Natural selection can do nothing until favourable variations chance to occur." [Darwin, 1859].) Mivart's contribution was to specify that constrained alterations in individual development, produced by reproductive peculiarities of the parents or changes in external conditions, are the necessary basis of evolution. As Pere Alberch was to say so well a century later, "In evolution, selection may decide the winner of a given game but development nonrandomly defines the players" (Alberch, 1980, p. 665). In many important details, Mivart also anticipated the contemporary concepts of macromutation and punctuated equilibria, the related notions that infrequent, relatively rapid, large-scale changes accurately portray the tempo and mode of evolution (reviewed by Stanley, 1981).

In sum, with special reference to our reviews of the ways in which the role of individual development in evolution has been conceptualized, Mivart's belief that changes relevant to evolution occur during ontogeny ("after or before birth, either in their embryonic, or towards . . . their adult, condition") is a most significant departure from previous thought on this topic. Lamarck, Darwin, and Haeckel all held that the only mode of evolutionary change was a change in adulthood, when a new stage got added on to the end of ontogeny (terminal addition). Mivart's insight on the importance of early ontogenetic change was not to be taken up again until much later (described in Chapters 8 and 9). Meanwhile, Haeckel's concepts of phylogeny and ontogeny got translated into the notion that there are two sources of input to individual development (nature and nurture), a dichotomy that became so firmly established (reified) that the study of genetics and embryology got separated, to the detriment of both fields. Those developments are recounted in the next two chapters, because they color the way we look at ontogeny and phylogeny down to the present day, impeding perception of the link between development and evolution.

◇ ◇ 6 ◇ ◇

Francis Galton: Nature Versus Nurture, or the Separation of Heredity and Development

Francis Galton (1822–1911) was Charles Darwin's first cousin and a great admirer of Charles's concept of natural selection as a major force in evolution. Galton studied humans and advocated selective breeding or nonbreeding among certain groups as a way of, respectively, hastening evolution and saving humankind from degeneracy. Galton coined the word *eugenics,* and its practice in human populations eventually resulted from his theories and the correlational studies of human traits by himself and Karl Pearson, among others.[1] Galton failed completely to realize that valued human traits are a result of various complicated kinds of interaction between the developing human organism and its social, nutritional, educational, and other rearing circumstances. If, as Galton found, men of distinction typically came from the upper or upper-middle social classes of nineteenth century England, this condition was not only a result of the selective breeding among "higher" types of intelligent and moral

Fig. 6–1. Francis Galton (1822–1911).

people but also was due in part to the rearing circumstances into which their progeny were born. This point of view is not always appreciated even today, that is, the inevitable correlation of social class with educational, nutritional, and other advantages (or disadvantages) in producing the mature organism.

Galton's dubious intellectual legacy was the sharp distinction between nature and nurture. Although he says in very contemporary terms, "The interaction of nature and circumstance is very close, and it is impossible to separate them with precision" (Galton, 1883/1907, p. 131). While it sounds as if Galton opts for the interpenetration of nature and nurture in the life of every person, in fact he means that the discrimination of the effects of nature and nurture is difficult only at the borders or frontiers of their interaction: "Nurture acts before birth, during every stage of embryonic and pre-embryonic existence, causing the potential faculties at the time of birth to be in some degree the effect of nurture. We need not, however, be hypercritical about distinction; we know that the bulk of the respective provinces of nature and nurture are totally different, although the frontier between them may be uncertain, and we are perfectly justified in attempting to appraise their relative importance" (1883/1907, p. 131).

Galton believed in measurement as the best empirical basis of scientific practice, so he measured not only traits such as the circumference of heads but even the efficacy of prayer. As far as the measurement of nurture goes, Galton held that the "furniture of a man's mind chiefly consists of his recollection and the bonds that unite them. As all this is the fruit of experience, it must differ greatly in different minds according to their individual experiences" (1883/1907, p. 131). Galton thus measured the kind and incidence of his own associations to various events, words, and happenings, and then divided up these recollections according to the period of his own individual development in which he believed they originated. The most pervasive associations were ones "whose first formation was in boyhood and youth," thus attesting to the importance and persistence of at least some effects of early experience into adulthood. Galton performed more or less the same sort of analysis for sentiments and found that

> the power of nurture is very great in implanting sentiments of a religious nature, of terror and of aversion, and in giving a fallacious sense of their being natural instincts. But it will be observed that the circumstances from which these influences proceed, affect large classes simultaneously, forming a kind of atmosphere in which every member of them passes his life. They produce the cast of mind that distinguishes an Englishman from a foreigner, and one class of Englishman from another, but they have little influence in creating the differences that exist between individuals of the same class. (1883/1907, pp. 154–155).

Would that Galton's list of "nurture" were more encyclopedic, but it begins and ends with the two categories of associations of ideas and associations of sentiments. As we shall see, Galton realized there are a myriad of environmental influences, but because the associations are what one could measure, that is what he presented to substantiate his argument. Elsewhere he wrote, "Nurture does not especially consist of food, clothing, education, or tradition, but it includes all these and similar influences whether known or unknown" (1875, p. 9).

Galton's entire scientific career revolved about the measurement of nature and nurture in one form or another, especially with respect to human achievement. His first book was *Hereditary Genius,* published in 1869, followed by *English Men of Science* in 1874, and *Inquiries into Human Faculty and Its Development* in 1883, which was a collection of his essays bearing on the assumption and methods involved in the measurement of nature and nurture.

Since we still retain, albeit unknowingly, many of Galton's beliefs about his concepts of nature and nurture, it will repay us to examine closely Galton's assumptions. Galton believed that nature, at birth, offered a potential for development, but that this potential or *reaction range,* as we would now call it, was rather circumscribed and very persistent.[2] "When nature and nurture compete for supremacy on equal terms . . . the former proves the stronger. It is needless to insist that neither is self-sufficient; the highest natural endowments may be starved by defective nurture, while no carefulness of nurture can overcome the evil tendencies of an intrinsically bad *physique,* weak brain, or brutal disposition" (1875, pp. 9–10, italics in original). One of the implications of this view is, as Galton wrote elsewhere, "The Negro now born in the United States has much the same natural faculties as his distant cousin who is born in Africa; the effect of his transplantation being ineffective in changing his nature." (1892, p. XXIV). The major conceptual error here is not merely that Galton is using his upper-middle class English or European values to view the potential accomplishments of another race, but it is rather that he has no factual knowledge of the width of the reaction range of African blacks—he assumes it not only to be inferior but to be narrow and thus without the potential to change its phenotypic expression. This kind of assumption is open to factual inquiry and measurement—it requires just the kind of natural experiment that Galton would have marveled at, and perhaps even enjoyed, given its simplistic elegance—namely, the careful monitoring and measurement of presumptively inbuilt traits *within generations* in races that have migrated to such different habitats, subcultures, or cultures that their genetic potential would be allowed to express itself in previously untapped ways. Thus, we can draw a line of increasing adult stature as Oriental groups migrate to the United States and substantially change their diet (recent review in Kimura, 1984, pp. 197–198). More important, we can measure the increase in IQ of blacks (within as well as between generations) as they move from the rural southern United States to the urban northeast, and its further increase the longer they remain in the urban northeast (Klineberg, 1935).

The reason I said that Galton might have reveled in these techniques of study, if not in the findings, is that they utilized the measurement procedures that he himself helped to pioneer in the areas of biology and psychology in the nineteenth century, namely, the concept of the normal distribution of traits and the frequency "error" of measurement, as it was then called. The concept is

> that the differences, say in stature, between men of the same race might *theoretically* be treated as if they were Errors made by Nature in her attempt to mould individual men of the same race according to the same ideal pattern. . . . One effect of the law may be expressed under this form. . . . Suppose 100 adult Englishmen to be selected at random, and ranged in the order of their statures in a row; the statures of the 50th and the 51st men would be almost identical, and would represent the average of all the statures. Then the differences, according to the law of frequency, between them and the 63rd man would be the same as that between the 63rd and the 75th, the 75th and the 84th, the 84th and the 90th. The intervening men between these divisions, whose numbers are 13, 12, 9, and 6, form a succession of classes, diminishing as we see in numbers, but each separated from its neighbors by equal grades of stature. . . . After some provisional verification, I applied this same law to mental faculties, working it backwards in order to obtain a scale of ability, and to be enabled thereby to give precision to the epithets employed. Thus the rank of first in 4,000 or thereabouts is expressed by the word "eminent." The application of the law of frequency of error to mental faculties has now become accepted by many persons, for it is found to accord well with observation. I know of examiners who habitually use it to verify the general accuracy of the marks given to many

> candidates in the same examination. Also I am informed by one mathematician that before dividing his examinees into classes, some regard is paid to this law. (1892, pp. XI–XII)

So, here we have the first so-called objective scale of measurement for mental faculties, variations of which are used everywhere today in mental testing, personality evaluation, and so forth. One prong of Galton's influence thus extends into mental measurement in psychology, and the other prong, with the substantial assistance of Karl Pearson in the early 1900s, reaches into biology as a momentous attempt to put evolution on a quantitative footing, both strands of Galton's influence persisting to the present day. But before considering the extraordinary impact of the latter in more detail, we need to return to Galton in 1869 to see the elegant simplicity with which he was to use his law of frequency of error to demonstrate the hereditary transmission of mental ability.

Like other nineteenth century thinkers, Galton's concept of hereditary was "like begets like." Thus, he was interested in determining whether the more eminent persons tended to beget more eminent persons, or, to say it another way, whether degrees of eminence were correlated in closely related human lines. In his own words, Galton was proposing

> to show . . . that a man's natural abilities are derived by inheritance, under exactly the same limitations as are the form and features of the whole organic world. Consequently, as it is easy . . . to obtain by careful selection a permanent breed of dogs or horses gifted with peculiar powers of running, or of doing anything else, so it would be quite practicable to produce a highly-gifted race of men by judicious marriages during several consecutive generations. (1892, p. 1)

Since Galton was to work with the historical record, he took it necessarily on assumption that "high reputation is a pretty accurate test of high ability." In this way, he was able to examine the relationships among a large body of men; namely, the judges of England from 1660 to 1868, the statesmen of the time of George III, and the premiers of England for the 100 years preceding his study. He then went on to examine the kin of the most illustrious commanders, men of literature and of science, poets, painters, and musicians. He also did the same for divines and scholars. Finally, he concluded his study by looking at what he considered to be the hereditary transmission of physical prowess, deducing same from the families of certain classes of outstanding oarsmen and wrestlers. In this pursuit, Galton obviously considered more than one grade of ability. His 1869 volume deals mainly with the small number "generally reputed as endowed by nature with extraordinary genius," perhaps 400 persons in all out of his sample of around 4000 gifted people. And these geniuses he found to be considerably interrelated. From our perspective today we would not be quite so certain that Galton was obtaining evidence for hereditary transmission, unless we included the social position, educational opportunities, and wealth that these kin also inherited. That criticism having been voiced, what we really need to know is that Galton's set of assumptions, his method of inquiry, and his statistical measures were, by way of Gregor Mendel, Karl Pearson, and Ronald Fisher, to eventuate in the mathematical concept of heritability, the mainstay of the study of population genetics in the twentieth century. The idea is, as Galton stated it, that one can select for virtually any trait in a population of individuals possessing the trait. The heritability coefficient of today is essentially a score (0 to 1.00) reflecting the degree of success in such selection (with higher correlations reflecting a greater degree of heritability). The heritability coefficient of today's breeder or geneticist always, even if quietly, assumes that the same rearing conditions hold between or among the generations being studied. So, properly interpreted, the heritability coefficient of today indicates how "true" some trait breeds given a more or less constant rearing environment. It is of great economic interest and importance to be able to predict that certain lines of dairy cattle will have above average (or below average) milk production, even if one is ignorant of the ontogenetic factors that must be present to bring about high (or low) milk production. But that is getting ahead of our story.

In terms of nineteenth century science, Galton's law of deviations from an average was an impressive mathematical contribution to psychology. With it he was able to show that what was true for aspects of physique or stature (as the French scientist Quetelet had done before him) also held for the distribution of psychological gifts and deficits within human populations. Given the interval classes above and below the mean, Galton could predict that about 250 men out of 1 million would be classified or graded as geniuses and about 250 men in 1 million would be classified or graded as idiots, with most of the population being judged mediocre or average, a somewhat smaller fraction above average and below average, and so on until the very small fraction of dolts and geniuses is obtained. Given the rather fixed or mechanical way the law always played out in relatively large populations, it was easy to see membership in the respective classes as being rather strictly determined or, one might say, predetermined. And for Galton it was rather narrowly predetermined by heredity. Galton's empirical case rested on "like begetting like" with far greater probability than would be expected by chance in families of eminent persons. He first documented this fact in *Hereditary Genius* in 1869 and followed it with a similar demonstration in *English Men of Science: Their Nature and Nurture*, published in 1874 in England (1875 in the United States). Galton was clearly interested in what he called "antecedents" of eminence in the life of the individual, but he was hampered in his quest because of the remote historical nature of his eminent persons. Eventually, he came upon what he believed to be a definitive way of delineating the relative importance of the contributions of nature and nurture in the ontogenetic development of the individual.

In Galton's *Inquiries into Human Faculty and Its Development*, originally published in 1883, he introduced the study of twins as the method par excellence of distinguishing between the effects of nature and nurture, or of "distinguishing between the effects of tendencies received at birth, and of those that were imposed by the special circumstances of their after lives" (1883/1907, p. 155). This was a "new method by which it would be possible to weigh in just scales the effects of Nature and Nurture, and to ascertain their respective shares in framing the disposition and intellectual ability of men" (p. 155). His idea was to discover twins who were judged to be very much alike and those that were judged to be "exceedingly unlike" early in life and then to study "how far their characters became assimilated under the influence of identical nurture, inasmuch as they had the same home, the same teachers, the same associates, and in every other respect the same surroundings." We must recall, as stated earlier, that it was Galton's supposition that nature would prevail over nurture. Here he believed was his opportunity to find scientific support for that supposition.

In Galton's time the distinction between identical twins (deriving from a single egg) and fraternal twins (deriving from two eggs) was understood. The first remark Galton makes is that discontinuity rules the psychological and behavioral aspects of twins—they do not obey the law of deviations. "Extreme similarity and extreme dissimilarity between twins of the same sex are nearly as common as moderate resemblances. When the twins are a boy and a girl, they are never closely alike: in fact, their origin is never due to the development of two germinal spots in the same ovum" (1883/1907, pp. 156–157).

Galton's study of twins was carried out by sending questionnaires to either the twins themselves or the near relatives of the twins. From a sample of eighty returns of cases of close similarity, he found thirty-five to be exceedingly close, not only in physical resemblance but in handwriting, voice, manner, and the like.

Galton does not flinch from interpretative difficulties in deciding when a late-developing difference between otherwise highly similar twins is due to nature or nurture. "It must be borne in mind that it is not necessary to ascribe the divergence of development, when it occurs, to the effect of different nurtures, but it is quite possible that it may be due to the late appearance of qualities inherited at birth, though dormant in early life, like gout" (1883/1907, p. 160). In this vein, Galton makes much over the numerous reports that twins often come down simultaneously, or nearly simultaneously, with the same noncontagious sickness, or that both develop a fear of heights at around age twenty, and so on (to be disputed in this century by Shields's [1962] study of identical twins). Also, the extremely close resemblance between certain twins includes their association of ideas. "No less than eleven out of the thirty-five cases testify to this. They make the same remarks on the same occasion, begin singing the same song at the same moment, and so on; or one would commence a sentence, and the other would finish it" (1883/1907, p. 165).

The acid test for Galton concerning the relative import of nature and nurture came when the thirty-five cases of closely allied twins reached early manhood or womanhood and went their separate ways. His analysis of the case reports revealed to him that

> in some cases the resemblance of body and mind had continued unaltered up to old age, notwithstanding very different conditions of life; and they showed in the other cases that the parents ascribed such dissimilarity as there was, wholly or almost wholly to some form of illness. . . . In only a very few cases is some allusion made to the

> dissimilarity being partly due to the combined action of many small influences, and in none of the thirty-five cases is it largely, much less wholly, ascribed to that cause. (1883/1907, p. 167)

Galton concludes,

> The twins who closely resembled each other in childhood and early youth, and were reared under not very dissimilar conditions, either grow unlike through the development of natural characteristics which had lain dormant at first, or else they continue their lives, keeping time like two watches, hardly to be thrown out of accords except by some physical jar. Nature is far stronger than Nurture within the limited range that I have been careful to assign the latter. (1883/1907, p. 168)

Galton then turns his attention to the other, to his mind, more important groups of twins, "those in which there was great dissimilarity at first, and will see how far an identity in nurture tended to assimilate them." He had a total of twenty cases "of sharply-contrasted characteristics, both of body and mind, among twins." He then goes on to cite thirteen quotations from parents or other close observers, several of whom actually say they believe the dissimilarity to reflect innate differences between the twins. Galton says that in none of the twenty cases is there any evidence of the "originally dissimilar characters having become assimilated through identity of nurture."

Having earlier in the same book argued for the pervasiveness and relative permanence of early associations and sentiments into adulthood, Galton notes that the seeming noneffect of nurture on different natures would seem to be at variance with that interpretation. He ends his study of twins, lamely, I am afraid, with the example of the young cuckoo who is reared exclusively in the nests of other species of birds. The cuckoo does not adopt the song or otherwise identify socially ("imprint") to its foster parents, but later finds its own species to mate with, sings a species-typical song, etc.

It remains to add that the same conceptual problems face us today as faced Galton and his colleagues in trying to understand the cuckoo as well as the similarities and dissimilarities among sets of twins. Today we are inclined to say that asking which is more important, nature or nurture, is not only an inappropriate question and a bogus starting point for an analysis of the ontogenetic development of the individual, but such an orientation actually provides an active stumbling block to developmental analysis. As my good friend Robert Cairns is wont to say, "Developmental analysis begins where the nature–nurture debate ends." This matter will be explored further in the next chapter on the conceptual impediments to the beginnings of the causal-analytic experimental approach to embryological development in the late nineteenth century.

◇ ◇ 7 ◇ ◇

August Weismann, Wilhelm Roux, Wilhelm His, and Hans Driesch: An Abortive Attempt to Understand Heredity Through an Experimental Approach to Embryonic Development

Galton's concept of nature in the nature–nurture dichotomy was instantiated or embodied in the unfolding of the fertilized germ during the course of embryonic development. In this conception, nurture, to the extent it operated at all, began to exert its major influence after birth. This was a very common or widespread scientific belief in the late 1800s. In fact, remnants of these beliefs persist to the present day in the idea that nature and nurture (or genes and environment) make separate contributions to individual development, and that the most formative force in early development is nature or genetics, with nurture or the environment becoming more influential after birth or even later in individual development. The experimental embryology of the late 1800s was originally addressed to an analytic understanding of the role of nature or the germ in producing the individual during the course of embryonic development. Unfortunately, as will be described in this chapter and the immediately succeeding ones, the analytic investigations failed to support the necessarily primitive concepts of heredity of the day, those concepts of heredity nevertheless went unchanged, and the respective study of heredity and development went their separate ways to the detriment of both.

Weismann's Germ Plasm Theory of Heredity

August Weismann's (1834–1914) germ plasm theory of heredity held that the nucleus of the fertilized egg or germ contained in it all the necessary "information" for the construction or assembly of the organism.[1] More specifically, Weismann, Wilhelm Roux (1850–1924), and many other scientists of the late nineteenth century believed the germ to be a highly complicated structure whose various parts ("determinants") corresponded to all the organs of the future organism. As noted by Hans Driesch (1908/1929), Weismann's idea was only a little less crude than the earlier belief in preformation. Since the epigenetic character of individual development was accepted as a fact by this time (the late nineteenth century), Weismann, Roux, and others saw the functional expression of heredity in the individual as the various components of the germ plasm giving rise successively to various organs during the course of embryonic development. The test of this idea of heredity and the germ plasm was in fact the object of one of the first experiments in what was to become the science of experimental embryology in the hands of Wilhelm His (1831–1904), Wilhelm Roux, and Hans Driesch (1867–1941).

Experimental Tests of Weismann's Theory of Heredity

In 1888, in an experiment that launched the truly experimental study of embryology, Wilhelm Roux, using a hot needle, killed one of the two existing cells after the first cleavage stage in a frog's egg

Fig. 7–1. August Weismann (1834–1914).

Fig. 7–2. Wilhelm Roux (1850–1924).

and observed the development of the surviving cell. Weismann's theory held that one-half of the hereditary determinants would be in each cell. And, indeed as called for by the theory, a half embryo resulted from Roux's experiment! However, a variation of the same experiment was subsequently carried out by Hans Driesch three years later, in 1891, with rather different results. Using the fertilized eggs of sea urchins, Driesch separated the two cells after cleavage by shaking them completely loose from one another and observed the

Fig. 7–3. Hans Driesch (1867–1941).

subsequent development of the single cell. From this single cell there arose an *entire* sea urchin, although diminished in size. This result was incompatible with any version of the Weismann–Roux theory of individual heredity and it also went against the related concept of preformation in an experimental way that had never before been accomplished.

As recounted in his book, *The Science and Philosophy of the Organism* (1908/1929), Driesch went on to manipulate the early cell divisions of the sea urchin in various ways and always obtained an entirely formed organism. In addition, he fused together two sea urchin eggs at the later blastula stage and produced a single giant organism from this fusion, rather than Siamese twins. That experiment was later repeated by Mangold in an amphibian with the same result, so Driesch's experimental findings were not attributable to a difference between the action of heredity in invertebrate (sea urchin) and vertebrate (frog) organisms. In fact, if Roux's original frog experiment is repeated and the separated cell is turned over, it develops into a whole frog embryo—it is only when left as is that it develops into a half embryo. Thus, the capacity to develop into a whole embryo was always present; the condition or circumstances of development determined whether the resulting phenotype would be a half- or wholly-formed organism.

In the 1890s, then, embryologists such as Roux, His, and Driesch were finding that the normal or typical outcomes of embryonic development were in part determined by the presence of normal or typical circumstances of development. Of the highest conceptual and theoretical importance, it was being learned that the potential or possibilities of morphogenesis or organogenesis exceeded whatever was actually obtained in individual outcomes. That is, as stated by Driesch (1908/1929, p. 53), the actual "fate" of a part of the body need not be identical with its possible fate in the sense that "there are more morphogenetic possibilities contained in each embryonic part than are actually realized in a special morphogenetic case." Thus, "there are more morphogenetic possibilities contained in each part than the observation of the normal development can reveal."

Driesch coined two terms to refer, respectively, to what is actually achieved in embryonic development (*prospective value*) and what might be possible (*prospective potency*). The goal of the science of embryology is summed up thusly, "If at each point of the germ something else *can* be formed than actually is formed, why then does there happen in each case just what happens and nothing else?" (Driesch, 1908/1929, p. 54).

Haeckel's Resistance to the New Experimental Embryology

Driesch was building on the groundwork for an *experimental* science of embryology, which had been in the making ever since Wilhelm His, in 1875, published a popular monograph with the title *Our Body and the Physiological Problem of Its Origin*. Twenty

Fig. 7–4. Wilhelm His (1831–1904).

years later, His's book was followed by Wilhelm Roux's frank call for a science of "developmental mechanics" (*Entwicklungs-mechanik*), in which the mechanics of embryonic development were to be examined by experimental means without recourse to the then-dominant biogenetic law. Ernst Haeckel immediately foresaw the threat of developmental analysis to his idea that "phylogeny is the mechanical cause of Ontogeny." Here is how Haeckel greeted these fresh analytic insights:

> Only the total want of critical ability and comprehension of the real problems of the History of Evolution can explain the fact that many people for a time regarded the strange fancies of . . . His as a great gain.
>
> Quite recently, however, His' . . . books on Ontogeny, which had previously ranked as the most perverted and unfortunate of the larger works on this science, have been eclipsed, in that respect, by a ponderous work. (Haeckel, 1897, vol. 1, pp. 64–65)

His, Roux, and Driesch were advocating that the morphology of animals was determined by proximal causes of ontogeny such as the protoplasm of the fertilized egg and the internal and external influences on its development, rather than accepting Haeckel's phylogeny as the mechanical cause of ontogeny via a preformationistic concept of heredity in which development was not seen as a dynamic and constructive process but merely as a realization of preexisting rudiments. Haeckel criticized His specifically for advocating "an entirely new view of the evolution [read development] of the body of Vertebrates, according to which the rudiments of the body of the vertebrate does not consist solely of the two primary germ-layers" (Haeckel, 1897 vol. 1, p. 64). As late as 1888, His was having to say:

> My attempts to introduce some elementary mechanical or physiological conceptions into embryology have not generally been agreed to by morphologists. To one it seemed ridiculous to speak of the elasticity of the germinal layers; another thought that, by such considerations, we "put the cart before the horse"; and one more recent author states, that we have better things to do in embryology than to discuss tensions of germinal layers and similar questions, since all explanations must of necessity be of a phylogenetic nature. This opposition to the application of the fundamental principles of science to embryological questions would scarcely be intelligible had it not a dogmatic background. No other explanation of living forms is allowed than heredity, and any which is founded on another basis must be rejected. The present fashion requires that even the smallest and most indifferent inquiry must be dressed in a phylogenetic costume, and whilst in former centuries authors professed to read in every natural detail some intention of the *creator mundi,* modern scientists have the aspiration to pick out from every occasional observation a fragment of

the ancestral history of the living world. The task of reading the chapters of this history seems to be as easy as collecting specimens of plants and animals, or of making microscopical preparations. The last principles of creeds and of theories are introduced into every empirical inquiry, and the danger is overlooked that even the best established theories frequently put a veil on the eyes of the observer, and interfere with the impartiality of his observations.

I should be the last to discard the law of organic heredity, or to deny the immense progress that biological science has made, by introducing this grand conception into the horizon. Questions of phylogeny will be for long of the utmost importance, and of the greatest interest in biology; but the single word "heredity" cannot dispense science from the duty of making every possible inquiry into the mechanism of organic growth and of organic formation. To think that heredity will build organic beings without mechanical means is a piece of unscientific mysticism. . . .

By comparison of different organisms, and by finding their similarities, we throw light upon their probable genealogical relations, but we give no direct explanation of their growth and formation. A direct explanation can only come from the immediate study of the different phases of individual development. Every stage of development must be looked at as the physiological consequence of some preceding stage, and ultimately as the consequence of the acts of impregnation and segmentation of the egg. . . .

Physiological considerations in morphology are far from interfering with phylogenetic inquiries, rather will the phylogenetic worker find in them a mighty help in his efforts. He has only to open his eyes to the actual processes of life and development. The operations which nature performs under our eyes cannot be different in principle from her processes in remote periods; and a good notion of the actual natural processes may, even for phylogenetic purposes, be much more useful than rigid morphological diagrams, abstracted by merely logical operations. (His, 1888, pp. 294–295)

Potential and Interaction as the Chief Meta-Theoretical Concepts Emerging from the Embryology of the Late Nineteenth Century

We have already seen how the various experimental manipulations of the early embryo typically caused some different outcome of development, thus giving rise to the notion that the heredity of the organism involved a reaction potential, bits and pieces of which were revealed depending upon the specific interactive influences that were allowed, or made experimentally, to operate during embryonic development. The range of the embryo's potential was seen also as changing during the course of development. Driesch's experiments showed that the potential became narrower with each step or stage of embryonic development, so that the early embryo had the greatest range of potential and the later embryo had the least potential in the sense of realizing alternative pathways and outcomes. The process of development itself was a limiting or delimiting factor.

There were also some nonsalutory concepts that emerged from this early era of experimental embryology, namely, the concepts of *self-differentiation* and *dependent differentiation*. These two terms were coined by Roux as a consequence of his half-embryo experiment, which he believed erroneously to be an outcome of self-differentiation, implying an independent or noninteractive outcome, in contrast to dependent differentiation where the interactive component between cells or groups of cells was necessary to, and brought about, the specific outcome. The concept of self-differentiation is akin to the concept of the innate when the term is applied to an outcome of individual development. Roux, himself, gave up the self- and dependent-differentiation dichotomy as he came to accept that his half-embryo result itself was the outcome of a particular interaction or influences, but, unfortunately, the concepts lived on in experimental embryology in disguised form as *mosaic development* versus *regulative development*. In the latter, the embryo or its cells are seen as developing in relation to the milieu, whereas the former is understood as a rigid and narrow outcome fostered by self-differentiation or self-determination, as if development were noninteractive. Here is the way W. K. Brooks (1902, pp. 490–491) expressed concern about the notion of self-differentiation:

> A thoughtful and distinguished naturalist tells us that while the differentiation of the cells which arise from the egg is sometimes inherent in the egg, and sometimes induced by the conditions of development, it is more commonly mixed; but may it not be the mind of the embryologist, and not the natural world, that is mixed? Science does not deal in compromises, but in discoveries. When we say the development of the egg is inherent, must we not also say what are the relations with reference to which it is inherent?

Experimental Embryology and the Germ-Plasm Theory of Heredity

In light of the findings of the new experimental embryology, August Weismann devoted his Romanes Lecture of 1894 to *The Effect of External Influences upon Development.* In that lecture, Weismann takes cognizance of Roux's notion that embryonic cells are adaptable (malleable) during the course of development and that the basis for these adaptations is, as Roux contended, the principle of Darwin's natural selection operating at the level of parts of the organism:

> Just as there is a struggle for survival among the individuals of a species, and the fittest are victorious, so also do even the smallest living particles contend with one another, and those that succeed best in securing food and place grow and multiply rapidly, and so displace those that are less suitably equipped. The three factors in the process of selection—variability, heredity, and struggle for existence—are all present. Processes of selection must thus take place amongst every kind of unit within the organism,—not only in cells and tissues, but also in the smallest conceivable living particles, which I have called "biophors." Everywhere equivalent parts are contending one with another, and everywhere it is the best that prevail. We can describe this process as *intra-individual* selection, or more briefly, as *intra-selection.*
>
> It is impossible for me to give an exhaustive account of Roux's argument here. . . . But there is one point I must not leave unnoticed: namely that relating to the cause which gives the advantage to one particle over others, and the consequent possibility of a struggle. This cause is to be sought in the relative power of reaction to a definite stimulus, and in the fact that a functional stimulus strengthens an organ. Just as the contraction of a muscle strengthens it, so also is every other histological element better nourished when acted on by the specific stimulus to which it is adapted. The varied sensitiveness for specific stimuli here has a similar result to that which follows in the case of the individual possessing certain advantages which make it victorious in the struggle with other individuals. In whatever part of the organism a definite stimulus is at work, there will necessarily be an increase of those elements that are most susceptible to this stimulus and are excited to the highest degree of activity by it. Thus elements which are stimulated to growth and increase by tension and pressure, necessarily accumulate and arrange themselves in the direc-

> tion of the stimulus in the parts where these forces act most strongly upon them. The arrangement of the spongy tissue of bones and of the complicated felting of the connective tissue in the dolphin's fin, as well as the marvellously suitable form and direction of blood vessels, are thus to be explained; and we may in general say that a similar explanation can be given to the various delicate adaptations of the tissues of the higher animals, all of which have the power of adapting themselves to the present circumstances of the organism. (Weismann, 1894, pp. 12–13)

Weismann thus adopts the highly interactive nature of embryonic development without making any essential changes in his preformative concept of the germ plasm. Before getting to that point in his Romanes Lecture, there are several choice quotations on what will later turn out to be preadaptation of the organism to the conditions of embryonic development: "Intra-selection effects the special adaptation of the tissues to special conditions of development in each individual . . . intra-selection effects the adaptation of the individual to its chance developmental conditions,—the suiting of the hereditary primary constituents to fresh circumstances. But these primary constituents themselves could only be produced by personal [i.e., natural] selection, and not by intra-selection" (Weismann, 1894, pp. 16–17).

Later on, Weismann actually uses his modification of Roux's intraselection to strengthen his preformative concept of the germ plasm:

> A complete harmony of the primary constituents can . . . never exist in the germ-plasm of sexually produced individuals; for this germ-plasm is always composed of two distinct halves. If, at any rate, those are right who agree with Darwin, Galton, de Vries, and myself in believing in a preformative arrangement of the germ-substance—that is, in a germ-substance composed of primary constituents (*Anlagen*)—it follows that in every act of fertilization very different primary constituents of corresponding parts, derived from father and mother, must meet in the germ. . . . The struggle of parts unequal in strength—that is, unequal in susceptibility to stimulus—must be the cause which brings about the mingling of parental primary constituents, and prevents the occurrence of monsters with parts ill adapted to one another. (Weismann, 1894, pp. 20–21)

Weismann then moves—inevitably, I think, given his most basic preformative assumptions—to posit a very fixed, wise, and unrealistic heredity, in which every eventuality has somehow been anticipated:

> The supposition of the whole activity of intra-selection presupposes the specific sensitiveness of the various primary constituents and of the units of smaller or larger groups of these; and this sensitiveness can naturally only have arisen through ordinary selection of individuals, owing to variation of the germ. For it is hereditary in this case just as in that of plants, the sensibility of which—(geotropic, heliotropic . . .)—dominates their whole growth. All these reactions of the organism to external influences are thus to a certain extent prearranged and provided for long in advance. (Weismann, 1894, pp. 21–22)

Thus, as stated before, in this viewpoint there is always a complete separation of ontogeny and phylogeny, and it is not appreciated that changes in embryonic development are at the basis of evolution or phylogeny, as was so clearly expressed by Mivart in an earlier chapter. Weismann says, "*The disappearance of a typical organ is not an ontogenetic but a phylogenetic process*" (1894, p. 35, italics in original).

Quite fittingly, Weismann ends his Romanes Lecture by once more acknowledging the all-sufficiency of natural selection in evolution:

> The facts here brought forward, as well as the interpretation of them at which I have arrived, confirm me more strongly than ever in the belief that selection is the all-sufficient principle on which the development of the organic world has been guided on its course. (1894, p. 54)

Puzzle of Heredity Becomes Dissociated from Problem of Individual Development

As we now believe that every cell in each organ of the body contains the same DNA, it will be understood immediately that individual development cannot consist merely of an unfolding of preexisting parts. Development proceeds from lower or simpler to higher or more complex levels of organization as, for example, in DNA giving rise (through messenger RNA) to proteins, which eventually become fully differentiated cells in each organ of the body (DNA → RNA → proteins → differentiated cells). Since the fully differentiated cell is not in the DNA, nor does the DNA or RNA give rise to a fully differentiated cell, it is the events that occur at the higher levels of organization above DNA and RNA (supragenetic events) that somehow bring about, and are thus required to fully explain, individual development. In the 1900s experimental embryology became devoted almost exclusively to the study of such higher level interactions. As we can now say with confidence, when the study of developmental genetics eventually becomes reunited with experimental embryology, we will come to understand heredity through an analysis of individual development.

The early pioneers of experimental embryology realized that there was a potential for understanding heredity by taking a causal-analytic approach to the study of embryonic development. Rather than move toward that potential, however, there was instead an ever-increasing divergence in the study of heredity and embryological development, such that, with the rediscovery of Mendel's work in early 1900, the geneticists would become uninterested in development and most embryologists would forsake the study of heredity.

◇ ◇ 8 ◇ ◇

Karl Pearson Versus William Bateson: The Foundation of the Quantitative Study of Heredity, or Genetics without Individual Development

The Darwinian theory of evolution by natural selection hinges entirely on the notion that favorable variations are preserved and unfavorable ones eliminated, so that a new species eventually arises very gradually, in slow steps, over the course of innumerable generations. Variation is thus the essential basis of evolution. Without variation natural selection would have no material on which to operate. A second essential ingredient of Darwinian evolution is that there must be hereditary mechanisms by which favorable variations can be transmitted to subsequent generations. As we realize now, it is the *capacity* to develop the favorable variations in the species-typical rearing environment that is transmitted between

Fig. 8–1. William Bateson (1861–1926).

generations. Thus, whether or not an organism is capable of a specific variation can only become apparent during the course of individual development. Consequently, genetics and embryology are essential to the understanding of the variations necessary for evolution. In the early 1900s the respective studies of genetics and em-

Fig. 8–2. William K. Brooks (1848–1908), unacknowledged mentor of William Bateson.

bryology became dissociated, because the geneticists ignored embryology and the embryologists ignored genetics.

William Bateson: The Origin of Variation

From my own reading, it has become apparent that the few scientists of the early twentieth century who strived to integrate the areas of genetics and embryology in the study of developmental genetics were intellectually isolated from their colleagues. The best example of such a person is William Bateson (1861–1926), the English biologist who coined the term *genetics*. He started his career with an interest in finding the likeliest ancestor of vertebrates. That took him to America in 1883, where he studied with William K. Brooks (1848–1908), an embryologist at the newly founded Johns Hopkins University. At the time, Brooks was in the throes of publishing his book *The Law of Heredity: A Study of the Cause of Variation, and the Origin of Living Organisms.* Although Bateson seems not ever to have acknowledged Brooks's influence, he (Bateson) subsequently turned his full attention, and devoted his whole career, to the study of variation. Bateson's intellectual isolation stemmed from the fact that he came to believe that natural selection was inadequate to the origination of species, echoing Mivart's idea that what was selected was born fit, so the problem was to understand the origin of variation. Again echoing Mivart, Bateson believed that the variation giving rise to new species was discontinuous, and his first book was titled *Materials for the Study of Variation, Treated with Especial Regard to Discontinuity in the Origin of Species,* published in 1894. In essence, Bateson saw species as being distinct entities that did not intergrade, so for him an understanding of evolution involved an understanding of discontinuous or saltational variations.

Karl Pearson: The Measurement of Continuous Variation

The dominant evolutionary thinking of the day was of course based on continuous variation shaped by natural selection. It was on this basis that Galton's intellectual successor, Karl Pearson (1857–1936), founded the scientific field of biometrics. The main aim of biometrics was to measure the continuous variation of traits within populations as the source or the potential for evolutionary change by natural selection favoring small differences between individuals. Biometrics later became a part of the field explicitly called *population genetics,* the dominant idea of which is that evolution is most likely caused by the changes in the frequency of genes in populations. This can be demonstrated by selecting certain individuals for breeding and increasing the appearance (or some aspect) of a particular phenotype in the population. While the genotypic and phenotypic composition of an interbreeding population can be so managed by artificial selection (as, for example, bringing about increased egg production in fowl or increased milk production in dairy cows), these changes are not considered of themselves to represent evolution but only to demonstrate the power of selective breeding to alter phenotypic expression. Thus, selective breeding provides

Fig. 8–3. Karl Pearson (1857–1926), founder of biometry.

indirect support for natural selection as a potential mechanism of evolution. When we later turn to a description of Ronald Fisher's genetic theory of natural selection, these ideas will be explored further.

Fig. 8–4. Gregor Mendel (1822–1884).

Rediscovery of Gregor Mendel's Research on Heredity in Certain Varieties of Garden Peas

Other than "like begets like," it seemed that nothing was known of the mechanisms of heredity during the nineteenth century. In actuality, however, Gregor Mendel's (1822–1884) work was published in 1865 and went unnoticed until 1900 when it was rediscovered by three independent geneticists working on related problems. Up to this time, the common notion was that characteristics of the parental organisms were fused or blended in their progeny. "Blending inheritance" was a stumbling block to the acceptance of natural selection by some scientists, because it would "swamp out" the effects of novelty and thus act against evolution. Galton, himself, at one time believed that selective breeding inevitably led to a regression toward the mean, which made it difficult to see how the selective breeding involved in natural selection could ever lead to the evolution of a new species, although it could lead to improvements by slowly shifting the mean. In that context, Mendel's work was especially significant because it showed the unblended, unitary character of inheritance, at least with respect to certain characteristics in the particularly well chosen varieties of garden peas with which Mendel worked. Mendel chose to work with twenty-two varieties of peas out of an initial thirty-eight varieties he tested for the first two years of his eight-year project. These varieties were especially well suited to the laws of segregation and independent assortment that Mendel discovered. Mendel studied seven dichotomous features of his peas:

1. Shape of seed (round or wrinkled).
2. Color of cotyledons (yellow or green).
3. Color of seed skin (gray or white).
4. Shape of seed pod (inflated or constricted).
5. Color of unripe pod (yellow or green).
6. Shape of flower arrangement (axial or terminal).
7. Length of peduncle (6 to 7 inches or ¾ to 1½ inches).

Very likely other features of the peas changed during the course of his experiments, but Mendel focused solely on these pairs of characters. He first crossed peas with these pairs of characters and observed in the first generation (F_1) that either one or the other character appeared, not a blend. The one that appeared he called dominant, the other recessive. The next step was to self-fertilize the cross-bred peas. In the F_2 he now recovered some with the recessive character in an approximate ratio of three dominant (75 percent) to one recessive (25 percent) phenotype. Mendel obtained these results with each of the seven pairs of characters. He now self-fertilized the F_2 peas. In the F_3 generation, the offspring of the recessive peas (25 percent of the population) remained recessive, and in subsequent generations never showed the dominant character again (i.e., they were pure recessives). When Mendel self-fertilized the dominant peas (75 percent of the total population), he discovered that one-third bred true thereafter for the dominant character (and thus were pure dominants), while two-thirds gave the three-to-one ratio of dominant to recessive. Thus, the Mendelian genetic distribution gives twenty-five dominant, fifty dominant-recessive, twenty-five recessive, or 1DD, 2DR, 1RR, with the phenotypic expression being three dominant to one recessive. This is an example of the simplest possible hereditary mechanism, where you have one factor (gene) for one trait, and one trait of a pair of traits being dominant and the other recessive. In such a simple genetic or factorial system, the inheritance is discontinuous (all or none); tall or short, wrinkled or smooth, yellow or green, etc. In fact, however, most traits of most organisms are multifactorially or polygenically determined and *phenotypic variation is most often continuous*. Nonetheless, the results of Mendel showed that certain factors of inheritance behave in a lawful manner especially if one focuses on particular traits in well-chosen varieties. The biometricians incorporated the discontinuous Mendelian factors into their genetic system by asserting that the traits they were measuring were multifactorial and so gave continuous distributions with otherwise discontinuous or independently assorting genes. This sort of conception allowed the genes to be looked on as self-contained packets of inheritance that give rise to certain traits and can be passed on from generation to generation. The one gene–one trait idea conformed well with Weismann's preformationistic concept of different portions of the germ containing the rudiment for specific later-developing traits in the organism. The self-contained nature of the packets of inheritance fits well with the notion that the germ is impervious to influences from the somatic cells, which influence was a necessary component of the unacceptable Lamarckian transmission of inherited characters. It is important for us to realize that concepts concerning genes and the mechanism of heredity at this time were inferences based on observations of phenotypic development.

William Bateson versus Karl Pearson: Concern for the Origin of Variation in Individuals versus the Measurement of Variation in Large Groups

As noted earlier, William Bateson believed natural selection to be inadequate as a means whereby species became transformed. This flew in the face of the guiding assumption of Karl Pearson's biometrics, which grew out of Francis Galton's quantitative studies of heredity and his interest in improving the human species by selective breeding (eugenics). That is, biometrics assumed that evolution occurred by natural selection favoring small deviations from the normal; in other words, that the variation underlying evolution is continuous and that new species arise very slowly by the long continued operation of natural selection operating in a given direction. Bateson, on the other hand, following Mivart and Brooks, believed that evolution was based on a phenotypically and genotypically discontinuous change, and that such changes possibly occurred rather abruptly through a one-in-a-million longshot; that is, the rare occurrence of a favorable mutation that arises during the course of individual development. Thus, according to Bateson, the study of development, particularly the study of cases of aberrant or unusual development such as, for example, twinning, would reveal the mechanism of evolution. He makes this point very clearly in his *Problems of Genetics* (Bateson, 1913/1979).

Bateson did not, however, have any clear hypothesis on what might be happening during development that could favor the emergence of a novel organism. He could foresee how a "slip" during the process of cell division could result in the loss of genetic material in that organism, which, if it happened frequently enough, eventually would lead to a recessive variety in some subsequent generation (Bateson, 1913/1979, pp. 90–91), but he could not fathom the evolution of a dominant factor. Did it come from an internal change (again, some accident in nuclear or cell division), or was it caused by some external mutagen? Bateson had a very unusual belief about genetic variation itself, which, I think made him uninterested in chromosomes or the constituents of chromosomes (the genes themselves) as the mechanism of evolution. He did accept the fact that with selective breeding or cross-breeding, one is manipulating the available genetic variation to get a new variety, but he could not accept that such selection or cross-breeding would ever produce a new species. Since the fundamental assumption of biometrics and population genetics is that species are derived from selection and cross-mating of one kind or another, Karl Pearson and, later, Ronald Fisher were at loggerheads with Bateson's thinking throughout their careers. Bateson's unusual belief about genetic variation seems only to have surfaced publicly once, in his presidential address to the British Association in 1914:

> If then we have to dispense, as seems likely, with any addition from without we must begin seriously to consider whether the course of Evolution can at all reasonably be represented as an unpacking of an original complex which contained within itself the whole range of diversity which living things present. I do not suggest that we should come to a judgment as to what is or is not probable in these respects. As I have said already, this is no time for devising theories of Evolution, and I propound none. But as we have got to recognise that there has been an Evolution, that somehow or other the forms of life have arisen from fewer forms, we may as well see whether we are limited to the old view that evolutionary progress is from the simple to the complex, or whether after all it is conceivable that the process was the other way about. When the facts of genetic discovery become familiarly known to biologists, and cease to be the preoccupation of a few, as they still are, many and long discussions must inevitably arise on the question, and I offer these remarks to prepare the ground. I ask you simply to open your minds to this possibility. It involves a certain effort. We have to reverse our habitual modes of thought. At first it may seem rank absurdity to suppose that the primordial form or forms of protoplasm could have contained complexity enough to produce the diverse types of life. But is it easier to imagine that these powers could have been conveyed by extrinsic additions? Of what nature could these additions be? Additions of material cannot surely be in question. We are told that salts of iron in the soil may turn a pink hydrangea blue. The iron cannot be passed on to the next generation. How can the iron multiply itself? The power to assimilate the iron is all that can be transmitted. A disease-producing organism like the pebrine of silkworms can in a very few cases be passed on through the germ cells. Such an organism can multiply and can produce its characteristic effects in the next generation. But it does not become part of the invaded host, and we cannot conceive it

taking part in the geometrically ordered processes of segregation. These illustrations may seem too gross; but what refinement will meet the requirements of the problem, that the thing introduced must be, as the living organism itself is, capable of multiplication and of subordinating itself in a definite system of segregation? That which is conferred in variation must rather itself be a change, not of material, but of arrangement, or of motion. The invocation of additions extrinsic to the organism does not seriously help us to imagine how the power to change can be conferred, and if it proves that hope in that direction must be abandoned, I think we lose very little. By the rearrangement of a very moderate number of things we soon reach a number of possibilities practically infinite. . . . Let us consider how far we can get by the process of removal of what we call "epistatic" factors, in other words those that control, mask, or suppress underlying powers and faculties. (cited by his wife, Beatrice Bateson, 1928, pp. 292–293)

Thus, Bateson seems to have believed that all the necessary genetic variation has always been present since the beginning of life and that the process of evolution involves the loss of inhibitors that otherwise prevent the action of hitherto unexpressed genetic factors (Hutchinson & Rachootin, 1979). Accordingly, since the genes themselves are not the secret avenue to an understanding of evolution, Bateson looked to the disinhibiting changes in individual development for the solution of the evolutionary question.

For Karl Pearson and his successor, Ronald Fisher, on the other hand, small changes in gene frequencies brought about by selective mating cause enough new genetic variation to produce new species. As already stated, Bateson did not believe that changes of gene frequencies within a species would yield anything more than new varieties. In addition, he felt that since species occupy distinct places, it is very unlikely that two species could cross in nature to produce a new species, thus making biometric and population-genetic analysis irrelevant to evolution from Bateson's point of view.

The facts may, I think, fairly be summarized in the statement that species are on the whole distinct and not intergrading, and that the distinctions between them are usually such as might be caused by the presence, absence, or inter-combination of groups of Mendelian factors; but that they are so caused the evidence is not yet sufficient to prove in more than a very few instances.

The alternative, be it explicitly stated, is not to return to the view formerly so widely held, that the distinctions between species have arisen by the accumulation of minute or insensible differences. The further we proceed with our analyses the more inadequate and untenable does that conception of evolutionary change become. If the differences between species have not come about by the addition or loss of factors one at a time, then we must suppose that the changes have been effected by even larger steps, and variations including groups of characters, must be invoked.

That changes of this latter order are really those by which species arise, is the view with which de Vries has now made us familiar by his writings on the Mutation Theory. In so far as mutations may consist in meristic changes of many kinds [changes in the numbers of bodily segments] and in the loss of factors it is unnecessary to repeat that we have abundant evidence of their frequent occurrence. That they may also more rarely occur by the addition of a factor we are, I think, compelled to believe, though as yet the evidence is almost entirely circumstantial rather than direct. The evidence for the occurrence of those mutations of higher order, by which new species characterized by several distinct features are created, is far less strong, and after the best study of the records which I have been able to make, I find myself unconvinced. The facts alleged appear capable of other interpretations. (Bateson, 1913/1979, pp. 100–101)

For once, Bateson and the biometricians were able to agree on something—neither side found de Vries's mutation theory satisfactory, even though Bateson himself felt the necessity of some sort of developmental mutation for evolution to occur. It is instructive to see that de Vries and Bateson held the same ideas about variation; namely, that there are two kinds of variation that occur in species: normal, small chance variations among individuals and larger differences brought about possibly by mutation. The former never lead to a permanent change in the average characters of the species. De Vries held that

if by stringent selection among such variations the mean character of the race is for a time changed, removal of the selection will be quickly followed by a regression to the old mean of the species. *Mutation,* on the other hand, is a phenomenon which occurs intermittently, and has not been shown to obey any ascertained law of magnitude or of frequency. An individual which exhibits a mutation belongs already to a new species; and its offspring exhibit regression not to the old specific mean, but to a new one . . . without mutation, therefore, no new species can be established; when a mutation has occurred a new species is already in existence, and will remain in existence, unless all the progeny of the mutation are destroyed. The only influence which natural selection can exert upon the course of evolution is that due to the total destruction of species. The phrase "survival of the fittest," as describing a process of evolution, ought to be replaced by "survival of the fittest species." (translated by Weldon, 1901, p. 366)

De Vries's *mutationstheorie* thus posed quite a challenge to the biometricians, who, of course, were basing their theory of evolution on the action of selection on normal individual variation. The biometrician W. R. F. Weldon (1901) made a searching critique of de Vries's monograph, *Die Mutationstheorie,* which was published in 1901, and reported de Vries's experiments and observations on what he believed to be the origin of new species in plants. It is instructive to follow Weldon's critique because it lays bare certain of the guiding assumptions of biometry and population genetics, the quantitative study of heredity. De Vries had described a selection experiment on wheat that showed that the trait selected was lost after the selected variety was returned to its usual habitat or rearing environment. If biometric selection can establish a new species, it must do so by the selected variation being able to exhibit and maintain itself under various conditions, at least according to de Vries's understanding. The failure of selection to bring about a stable variation meant to de Vries that selection was inadequate to bring about evolution unless the new form "was born fit" through an abrupt, large-scale mutation. Here is the way Weldon handles de Vries's argument about the wheat example:

The race to which this wheat belonged originated in temperate Europe, but by selection among plants grown in Norway, near the northern limit of possible culture, a form was produced which ripened earlier, and had heavier seeds than the parent form. Seeds from this form, when sown in more southern countries, gave rise after a few generations to plants which resembled the [original] parent race. Here we have obviously to consider not only the cessation of selection, but the change in external conditions, as affecting the result. Again Professor de Vries himself shows that the number of supernumerary carpels in the fruit of *Papaver somniferum polycephalum,* produced by plants grown from seed of the same parental fruit, varies enormously (from 150 to one or two!) according to the amount of nutrition supplied during particular stages of growth; he says deliberately that the selection of plants with the greatest number of carpels is simply the selection of the best nourished individuals; and yet the reduction in the number of extra carpels after cessation of selection is quoted as proof that the results of selecting mere variations are unstable.

Now it cannot be too strongly insisted upon that every character of an animal or of a plant, as we see it, depends upon two sets of conditions; one a set of structural or other conditions inherited by the organism from its ancestors, the other a set of environmental conditions. There is probably no race of plants or of animals which cannot be directly modified, during the life of a single generation, by a suitable change in some group of environmental conditions.

The work of Dareste, Driesch, Herbst, and others has shown that some of the most normal and universal phenomena of animal development are each directly dependent for their occurrence upon a certain group of external conditions. . . . Until we know far more than we know at present about the relation between an organism and its environment, it is simply useless to discuss the stability of characters, whether "variations" or "mutations," except under environmental conditions which are as constant as we can make them during the period under discussion.

The characters which give their value to the improved races of wheat, and to many of our cultivated plants, are admittedly in large part the direct result of cultivation under special conditions; and in order to judge whether the effect of selection on such plants is permanent we must grow them without selection under the same carefully arranged conditions of nutrition as those adopted for the culture of the race during the operation of selection. The evidence as it stands gives little or no indication what their behaviour under such circumstances would be. (Weldon, 1901, pp. 367–368)

It was of utmost import to the biometric point of view that selection could bring about stable varieties, without the necessity of mutation, based only on the normally occurring small differences among individuals. It was Pearson's assumption that if there was sufficient variation in a population, the mean of each generation could be changed to slowly bring about evolution (or at least a new variety). While individual development was of great importance in Mivart's, Bateson's, and de Vries's points of view, it was the statistics of large populations or masses that counted in the biometric tradition:

> The starting point of Darwin's theory of evolution is precisely the existence of those differences between individual members of a race or species which morphologists for the most part rightly neglect. The first condition necessary, in order that any process of Natural Selection may begin among a race, or species, is the existence of differences among its members; and the first step in an enquiry into the possible effect of a selective process upon any character of a race must be an estimate of the frequency with which individuals, exhibiting any given degree of abnormality with respect to that character occur. The unit with which such an enquiry must deal, is not an individual but a race, or a statistically representative sample of a race; and the result must take the form of a numerical statement, showing the relative frequency with which the various kinds of individuals composing the race occur.
>
> As it is with the fundamental phenomenon of variation, so it is with heredity and with selection. The statement that certain characters are selectively eliminated from a race can be demonstrated only by showing statistically that the individuals which exhibit that character die earlier, or produce fewer offspring, than their fellows; while the phenomena of inheritance are only by slow degrees being rendered capable of expression in an intelligible form as numerical statements of the relation between parent and offspring, based upon statistical examination of large series of cases, are gradually accumulated. . . .
>
> It is almost impossible to study any type of life without being impressed by the small importance of the individual. In most cases the number of individuals is enormous, they are spread over wide areas, and have existed through long periods. Evolution must depend upon substantial changes in considerable numbers and its theory therefore belongs to that class of phenomena which statisticians have grown accustomed to refer to as *mass-phenomena*. A single individual may have a variation which fits it to survive, but unless that variation appears in many individuals, or unless that individual increases and multiples without loss of the useful variation up to comparatively great numbers—shortly, until the fit type of life becomes a mass-phenomenon, it cannot be an effective factor in evolution. The moment this point is grasped, then whether we hold variation to be continuous or discontinuous in magnitude, to be slow or sudden in time, we recognise that the problem of evolution is a problem in

> statistics, in the vital statistics of populations. Whatever views we hold on selection, inheritance, or fertility, we must ultimately turn to the mathematics of large numbers, to the theory of mass-phenomena, to interpret safely our observations. As we cannot follow the growth of nations without statistics of birth, death, duration of life, marriage and fertility, so it is impossible to follow the changes of any type of life without its vital statistics. (Pearson, 1901, pp. 2, 3)

Bateson's point of view about the importance of the biometrician's individual differences, as I stated earlier, is that they could not instigate or support evolution. The only kind of variation that could do so was, according to Bateson (1894), the variation that occurred between parent and offspring, showing that the offspring, for whatever reason, were differentiating away from the parental type. This is what Bateson considered to be discontinuous variation. Although he was never too clear or explicit in a detailed way, Bateson believed that evolution was initiated by some sort of developmental accident that resulted in a change or rearrangement (most likely, a deletion or disinhibition) of genetic material in the offspring during ontogeny. As we shall see, this view was never to be very popular, even in the twentieth century, as the quantitative biometric study of population genetics became a field unto itself, dissociated from a consideration of embryology and individual development.

◇ ◇ 9 ◇ ◇

Walter Garstang, Gavin de Beer, and Richard Goldschmidt: The Concept of Changes in Individual Development as the Basis for Evolution

The concept that changes in individual development are the basis for evolution was raised originally by St. George Mivart in his book *On the Genesis of Species* (1871). William Bateson also favored the idea but little was done to work out the details until Walter Garstang (1868–1949) and Gavin de Beer (1899–1972) delivered their respective coups de grâce to Haeckel's recapitulation doctrine, and, from another side entirely, Richard Goldschmidt (1878–1958) hypothesized that changes in early embryonic development would be necessary for evolution to occur. While Garstang and de Beer were interested in showing the importance of various kinds of ontogenetic changes to evolution generally, Goldschmidt, having become convinced of the impossibility of neo-Darwinian microevolution producing a new species, had come to the notion of a developmental macromutation as essential to the production of the large differences necessary for speciation.

Garstang's Critique of Recapitulation

Garstang's (1922) critique of recapitulation, which in effect stood Haeckel's biogenetic law on its head, had the positive effect of opening the door for the notion that all sorts of changes in ontogeny (not only additions at the end of development) were the basis of evolutionary change. The avowed purpose of Garstang's critique was to create a scheme other than Haeckel's for the theoretical understanding and explanation of the relations between ontogeny and phylogeny.

For Haeckel, phylogeny meant the phyletic (ancestor–descendant) line of succession of *adults*. In Haeckel's scheme the ontogenetic sequence in a descendant was caused by the phyletic sequence of adults that had preceded it: "Phylogeny is the mechanical cause of ontogeny." Garstang held that Haeckel's view overlooked the fact that the phyletic line of succession of zygotes (fertilized eggs), running more or less parallel with the adult sequence, was itself steadily diverging.

> Every elaboration of adult form, even of its degree of pliability under environmental influence . . . was preceded by a corresponding elaboration of zygotic structure, nuclear or cytoplasmic or both, determining, under suitable conditions, the form and character of the ontogenetic changes and their result. Through the whole course of evolution, every adult Metazoan has been the climax of a separate ontogeny or life-cycle, which has always intervened between adult and adult in that succession of forms which Haeckel terms "Phylogenesis." The real Phylogeny of Metazoa has never been a direct succession of adult forms, but a succession of ontogenies or life-cycles. (Garstang, 1922, p. 82)

Phylogeny is thus not the cause but the product of a succession of different ontogenies. Consequently, Garstang held that ontogeny does not recapitulate phylogeny: it creates it. In essence, what

Fig. 9–1. Walter Garstang (1868–1949).

Garstang did was to put von Baer's nonevolutionary generalizations concerning ontogeny into an evolutionary framework, as Figure 9–2 indicates.

This way of representing the relations between ontogeny and phylogeny is still a recapitulation of sorts. As von Baer had observed, related species show early ontogenetic similarities and diverge from one another as ontogeny proceeds. In putting von Baer's ontogenetic framework into an evolutionary context, Garstang was saying that

> ontogeny proceeds through successive *grades of differentiation* by which layers, tissues, organs, and parts together with ordinal, family,

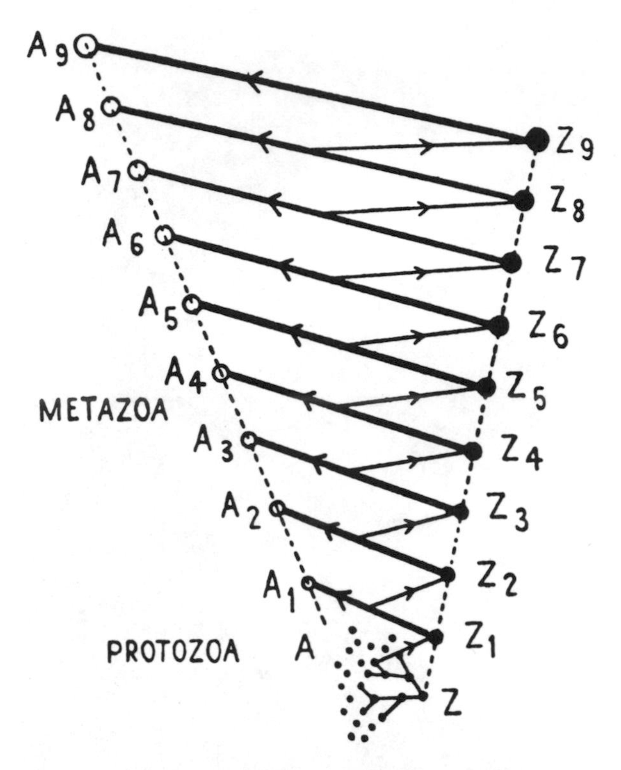

Fig. 9–2. The relationship of ontogeny to phylogeny in morphological evolution, according to W. Garstang. Z, Z_1, Z_2 and A, A_1, A_2, and so on refer, respectively, to zygotes and adult phenotypes in a particular phylogenetic lineage. In Garstang's view, mutation or genetic recombination gives rise to an atypical zygote with a corresponding altered ontogeny. The altered zygote and its altered ontogeny give rise to a novel adult phenotype (i.e., a new taxon). (From Garstang, 1922.)

generic, and specific characters, are more or less successively established. As differentiation increases, the combination of layers, tissue, organs, and parts exhibited at successive stages resemble more or less distinctively the combinations characteristic of successive *grades of evolution* represented in our schemes of phyletic classification. (Garstang, 1922, p. 84)

In Garstang's scheme, later evolving ontogenies are seen as modifications of earlier occurring ontogenies, as a consequence of changes in the zygote or fertilized germ. Thus, changes in ontogeny are the expression of an altered zygote and that expression is not merely one of terminal addition but represents changes in earlier stages of ontogeny.

> The ontogeny which first established the Coelenterate grade was the basis of a later ontogeny which established the Coelomate grade. The life-cycle was extended accordingly, but never by the simple addition of a substantial unit or stage, distinctively Coelomate, to the final adult stage of a Coelenterate ontogeny. A house is not a cottage with an extra story on the top. A house represents a higher grade in the evolution of a residence, but the whole building is altered—foundations, timbers, and roof—even if the bricks are the same. (Garstang, 1922, p. 84)

Garstang's mechanism of early ontogenetic modification was embryonic mutation. These mutations, he held, would be subjected continuously to selective tests of their physiological efficiency, and those that passed such tests would survive. In his usual succinct way Garstang stated that "age bears the buffets of the world, but youth regenerates it." This is most reminiscent of Mivart's contention that what survived selection was born fit. Other than "embryonic mutation," arising either from a presumptively changed zygote or a developmental accident of some sort, Garstang cannot give us any further hints of how the ontogenetic changes so critical to evolution originate. But Garstang unflinchingly faces the logical conclusion from his analysis, to the effect that the first metazoan was not produced by a metazoan but by "a genius among Protozoan zygotes." And, if that is not clear enough, he goes on to say, in the conclusion of his essay, "The first Bird was hatched from a Reptile's egg." Thus, in stark contrast to the way in which Darwin and Haeckel wanted embryology to bear witness for phylogeny, "Ontogeny is not an animated cinema show of ancestral portraits; but zygotes may be likened to conjurers playing the old tricks for the most part, and occasionally opening a surprise packet—nor do they always keep their novelties back until the end of the performance" (Garstang, 1922, p. 100).

Fig. 9–3. Gavin de Beer (1899–1972).

De Beer's Systematic Elaboration of Garstang's Critique of the Biogenetic Law

Gavin de Beer elaborated on Garstang's proposal in a very systematic way, beginning with his first book on the topic, *Embryology and Evolution,* published in 1930, and culminating with the third edition of *Embryos and Ancestors,* which appeared in 1958. De Beer was a first-rate scholar and acknowledged that much of his argument was based on the observations of von Baer, although he had to recognize that von Baer not only repudiated recapitulation but that he rejected a Darwinian sort of monophyletic evolution as well.

Before getting down to the task at hand, de Beer (1958) reminds the reader that the order in which characters appeared in phylogeny is not always faithfully reproduced in ontogeny. The example he gives is that teeth were evolved before tongues but in mammals now tongues develop before teeth. Thus, in the ontogenetic development of present-day mammals, the appearance of teeth has been retarded relative to the tongue, thus falsifying the argument that *because* a structure is formed early in embryonic development, *therefore* it must have appeared early in evolution. This, and many similar examples, serves as grist for de Beer's mill that *heterochrony* (alteration and reversal of ontogenetic sequences) is a major facet of evolution and not a troublesome "false testimonial" to phylogeny as Haeckel saw it. De Beer goes on to quote (rather mischievously one imagines) W. K. Gregory's (1925) remark that "If the biogenetic law was universally valid, it would seem legitimate to infer that the adult common ancestor of man and apes was a peculiar hemaphroditic animal, that it subsisted exclusively upon its mother's milk, and that at an earlier phylogenetic period the adult ancestor was attached to its parent by an umbilical cord."

As we saw in the chapter concerning the rise of experimental embryology, Haeckel's widely shared idea that phylogeny was the *mechanical* cause of ontogeny rendered the analysis of ontogeny superfluous—one was called upon merely to observe ontogeny, not to examine it experimentally, because the ancestral record read itself off like an internally driven ticker tape, completely impervious to its developmental surroundings. This influential concept of Haeckel's survives in disguised form today in the concept of "innate" in animal development. De Beer was one of the few writers, outside of the experimental embryologists, to realize the conceptual inadequacy of Haeckel's notion of the relation of heredity to ontogeny. Rather than the passive translation of phylogenetic causes into ontogenetic happenings, ontogeny in each generation is a consequence of the coaction of hereditary or genetic factors and many different local environing circumstances that determine the expression of the phenotype during the course of development. Normal or species-typical development is only inevitable if (1) the individual genotype is within the usual range of variation, (2) no developmental accidents happen (e.g., mutation), and (3) the usual environing conditions from cell to organism prevail. Thus, before getting into his analysis of the role of ontogeny in phylogeny, de Beer establishes the interactional or coactional nature of what he calls internal and external factors in embryonic development. He notes that ever since the Silurian period, about 300 million years ago, vertebrate animals have had two eyes, as is borne out by the fossil record. Thus, there must be a hereditary or genetic capacity to develop two eyes in vertebrates, and this capacity has been transmitted to every generation for a considerable period. "But these hereditary factors are not self-sufficient, for if a few pinches of simple salt (magnesium chloride) are added to the water in which a fish (*Fundulus*) is developing, that fish will undergo a modified process of development and have not two eyes, but one. . . . Countless similar examples might be given, but this one suffices to show that by themselves the internal and therefore transmitted factors are not able to "produce" a normal animal" (de Beer, 1958, p. 14). De Beer then goes on to cite others who have shown that internal and external factors cooperate in the production of all characters of an organism. He further exphasizes the distinction between the process of transmission of the internal factors from parent to offspring, and the process of production in the offspring of characters similar to those which were possessed by the parent. He cites E. S. Goodrich (1924), who said, "An organism is moulded as the result of the interaction between the conditions of stimuli which make up its environment and the factors of inheritance. No single part is completely acquired, or due to inheritance alone. Characters are due to reponses, and have to be made anew at every generation." De Beer's point is that the question, Are acquired characters inherited? has no meaning, for all characters of an organism are both inherited and acquired; "they would not be developed at all unless the organism possessed the requisite internal and inherited factors, and unless the external factors were sufficiently 'normal' to evoke the 'normal' developmental responses. A change in either the internal or the external factors will result in a departure from normal development."

By way of establishing his frame of reference for the understanding of ontogeny itself, as well as the relationship between ontogeny and phylogeny, de Beer makes the point that ontogeny is not merely an extrapolation into the future of a chain of events that occurred in past and previous generations:

> Each ontogeny is a fresh creation to which the ancestors contribute only the internal factors by the means of heredity. And this historic past is not the phylogenetic line of ancestral adults, but the line of the germ-plasm supplying the fertilized eggs for each and every generation. The action of the internal factors is to ensure that if the external factors are normal and do evoke any response in development and produce an animal at all, that animal will develop along the same lines as its parent. The internal factors are only a partial cause of ontogeny. (de Beer, 1958, p. 17)

Before describing the eight ways or modes in which changes in the timing of ontogenetic events bring about evolution, de Beer acknowledges our ignorance of the causes of such ontogenetic changes. Beyond the statement that mutation and genetic recombination are involved, the actual origin of evolutionarily significant changes in ontogeny is unknown. Since not only genes but external and environmental factors must be involved in such transformations, a truly comprehensive account of evolutionary changes will necessarily include the supragenetic components as well as the presumptive genetic changes attendant to evolution, but that is getting ahead of the story.

De Beer's major idea was that evolutionary change derived from ontogenetic changes in the timing of the various aspects of individual development in descendants as compared to ancestors. In this view, changes in timing (heterochrony) produce a different looking phenotype by the rate of growth of certain stages being either accelerated or retarded relative to other stages of development. De Beer's evolution via heterochrony was entirely compatible with D'Arcy Thompson's "laws of the growth of form" (1917), which took a von Baer-like basic body plan and changed the size and position of the parts relative to one another in geometric ways to show how one can, by this procedure, get a variety of animal forms, thus showing how evolution might proceed (Figure 9–4). It was also compatible with the then-prevalent idea of Goldschmidt (see later) and of Ford and Huxley (1927) that genes determine the rate at which events occur during embryonic development. It was a simple step to envision how genetic recombination or mutation

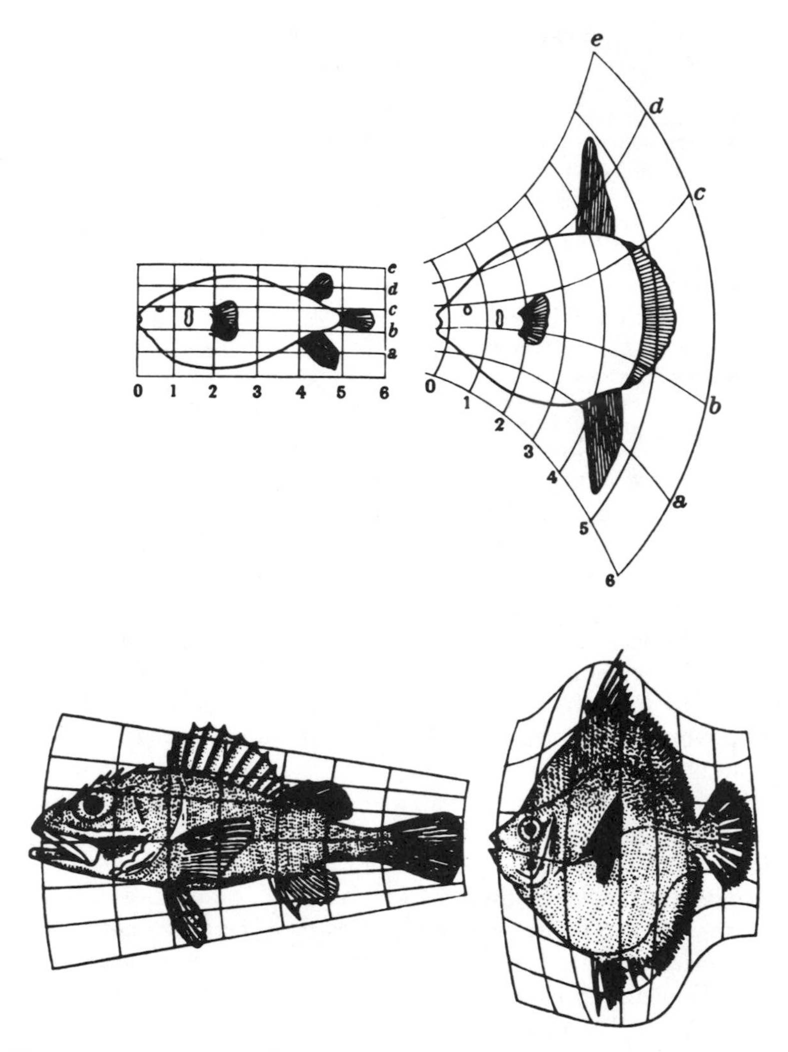

Fig. 9–4. Relatively small geometric transformations of a basic body plan could give rise to new morphologically defined species in a lineage of related forms (from D'Arcy Thompson, 1917). These outcomes could be accomplished by relative changes in the rate and/or duration of growth of various body parts during ontogeny. The various ways in which developmental changes in timing can be accomplished are shown in Figure 9–5.

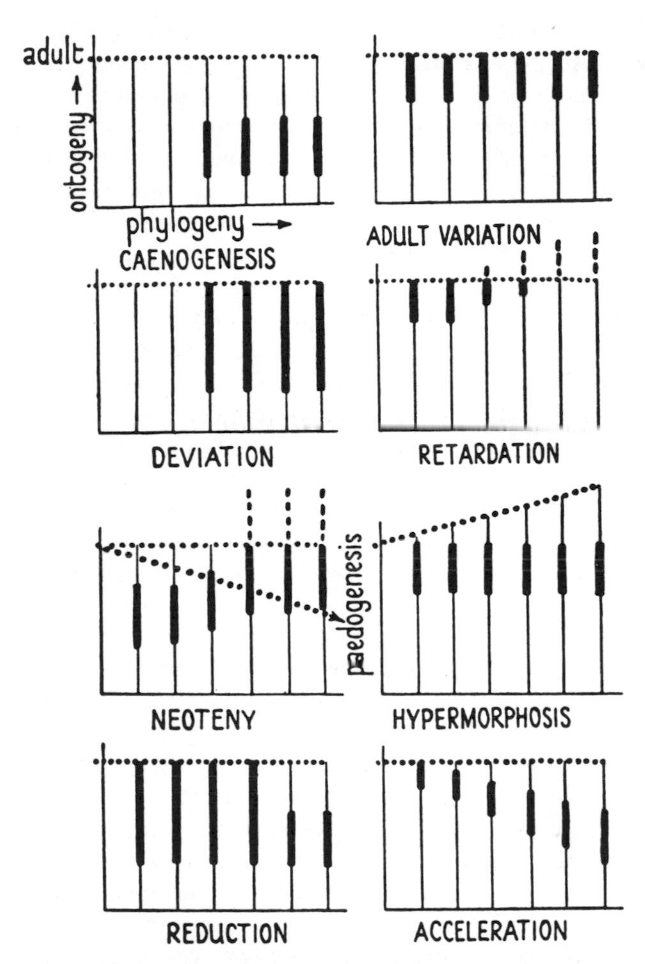

Fig. 9–5. Gavin de Beer's eight morphological modes describing how various known alterations of ontogeny could give rise to a new taxon. Dark bar indicates the temporal appearance of an ontogenetic novelty. (From de Beer, 1958.)

could effect changes in the timing and thus the expression of character in ontogeny.

De Beer formalized the eight morphological modes in which ontogenetic acceleration and retardation could produce evolution, as shown in the diagram in Figure 9–5.

Brief definitions and examples of certain of de Beer's modes follow: *caenogenesis* (also spelled kenogenesis, cenogenesis) refers to the introduction of a morphological novelty into the youthful stage of development, as in the abrupt evolutionary appearance of embryonic membranes (amnion, allantois) in reptiles, birds, and mammals. These are strictly embryonic adaptations necessary to the embryonic life of these three classes of organisms, the embryos of which then go on to diverge (show evolutionary *deviation*) in their adult form. *Neoteny* refers to the retention of ancestral embryonic or youthful stages in the adult stage of the descendant, a condition earlier called *foetalization* by Bolk (1926, and augmented by Montagu, 1962) in describing the most conspicuous evolutionary changes in humans. In plain English, these are cases of arrested development and were termed *paedomorphosis* by Garstang (1922). An example of neoteny in our own species, one which distinguishes us from all other mammals and probably all other vertebrates, is our retention into adulthood of the cranial flexure of the fetal period (eyes and nose facing at right angle to spine). This embryonic condition is exhibited by all other mammals, which then deviate such that the eyes and nose form a more or less continuous line with the angle of the spinal cord in the adult form. The retention of this embryonic feature in humans is accompanied by others, such as the retention of the fetal skull shape and nonopposable big toe, all of which are necessary to, or are correlated with, upright walking posture. De Beer was able to compile a very large number of presumptive instances of neoteny in a variety of species, thereby suggesting its probably wide utility as an evolutionary mode. He was quick to point out that when an adult descendant resembles the young stage of an ancestor, that is precisely the "reverse and opposite of that which would be required under the theory of recapitulation."

To avoid the possibly tedious recital of other examples of heterochrony offered by de Beer, I would urge interested readers to consult his short and concise monograph on this topic (de Beer, 1958). The main point is that de Beer was the first to propose a systematic account of the various ways development could be altered to produce evolutionary change, and he produced a veritable catalogue of examples of such changes, thereby strengthening his argument that developmental heterochrony was one of the principal means of evolutionary change. He accepted the conventional assumptions of the time, to wit, that mutations or genetic recombination through selective breeding could produce these large-scale heterochronic changes, the genes being viewed as determining the rates of activity of different embryonic processes in forming the organism. In his book on the topic of ontogeny and phylogeny, Stephen Gould (1977) has reworked de Beer's categories and once again cast heterochrony as a prime mover of evolution.

Richard Goldschmidt and the Concept of a Hopeful Monster

Richard Goldschmidt (1878–1958) began his scientific career early in the twentieth century by taking Darwin at his word as to the origin of species. (Darwin had held that varieties are incipient species.) Consequently, Goldschmidt experimentally subjected various subspecies of the gypsy moth to various environmental and selective breeding regimens, expecting thereby to eventually produce a new species, as predicted by Darwin and the neo-Darwinists of the early 1900s, especially the biometricians. Over the course of more than twenty-five years Goldschmidt performed these experiments on a number of subspecies for thousands of generations before giving up the idea that evolution occurred as a consequence of the selection and accumulation of a number of small variations. The best Goldschmidt could do was to produce more subspecies or races of the same species, never a new species. This discouraging result of more than a quarter of a century's labor led Goldschmidt to echo William Bateson's pessimism concerning the concepts of continuous variation and Mendelian genetics as providing a basis for evolution. Goldschmidt thus turned to the idea of a hopeful monster as the only way to get a new species or genus. This would involve a sudden, large-scale discontinuity, possibly arising in a single generation, and somehow perpetuating its novel genetic makeup for a sufficiently long period so that other similar hopeful monsters might eventually form a new group of animals. Goldschmidt (e.g., 1933) granted the rarity and unlikely nature of such a mutation but he saw it as the only real possibility for evolution to occur above the subspecific level: "The changes necessary for the formation of a new species are so large that the relatively small differences of the subspecies as a starting point would hardly count" (1933, p. 542).

From Goldschmidt's point of view, three things had to happen

Fig. 9–6. Richard Goldschmidt (1878–1958).

to produce a new species. The first two of these were, and are, widely shared assumptions:

> (1) The transformation of one species into another is possible only if permanent changes in the genetic make-up occur, and (2) if the changed forms stand the test of selection. . . . But there is a third point, often neglected, which lies, I think, at the basis of the whole problem, namely, (3) the nature of the developmental system of the organism which is to undergo evolutionary change. The appearance of a genetic form, whether we call it a species or a genus, which is to be considerably different from the ancestral forms, requires that a considerable number of developmental processes between egg and adult have to be changed, in order to lead to a different organization. Development, however, within a species is, we know, considerably one-tracked. The individual developmental processes are so carefully

> interwoven and arranged, so orderly in time and space, that the typical result is only possible if the whole process of development is in any single case set in motion and carried out upon the same material basis, the same substratum and under the same control by the germ plasm or the genes. From this it follows that changes in the developmental system leading to new stable forms are only possible as far as they do not destroy or interfere with the orderly progress of developmental processes. Of course, everybody knows that this is the reason why most mutations are lethal. But not everybody keeps in mind that here also is touched one of the basic points of the problem of evolution. The nature and working of the developmental processes of the individual then should, if known, permit us to form certain notions regarding the possibilities of evolutionary changes. (Goldschmidt, 1933, p. 543)

In his stress on the integrity and centrality of the developmental process, Goldschmidt echoes Mivart's views, as he does on the orthogenetic (directional) character of evolutionary (i.e., developmental) change and the need for a macromutation to spur such evolutionary change. Goldschmidt is able to go beyond Mivart by specifying that the most viable mutations will affect the end process of embryonic differentiation and thus will little alter the basic appearance and physiology of the extant species. What is required for evolution is the rarely successful mutational changes that act upon earlier developmental processes and, thus, most surely produce a monster.

> The most probable mutational change with a chance to lead to a [functional] organism is a change in the typical rate of certain [early] developmental processes. Of course, in most cases such a shift of a partial process would lead to the production of monstrosities. . . . But we must not forget that what appears today as a monster will be tomorrow the origin of a line of special adaptations. The dachshund and the bulldog are monsters. But the first reptiles with rudimentary legs or fish species with bulldog-heads were also monsters . . . [Thus,] I cannot see any objection to the belief that occasionally, though extremely rarely, [an early mutational change in the rate of developmental processes] may act on one of the few open avenues of differentiation and actually start a new evolutionary line . . . then an avenue would be open to considerable evolutionary change with a single basic step, provided that the new form could stand the test of selection, and that a proper environmental niche could be found to

> which the newly formed monstrosity would be preadapted and where, once occupied, other mutations might improve the new type. [Goldschmidt, 1933, pp. 544–545)

As an example of the kind of macromutation he had in mind, Goldschmidt cites the abrupt and immense change of the hyomandibular bone, which, in fishes, connects the jaws to the skull, but which, in amphibians, serves as an auditory ossicle inside the cavity of ear. This developmental (evolutionary) change was accompanied by the appearance of a tympanic membrane, thus allowing the hearing of airborne sounds in the first terrestrial creatures. In both cases of change a slow transformation by the accumulation of small advantageous mutations is unimaginable. Thus, to summarize to this point, Goldschmidt felt that the larger steps in evolution are "to be understood as sudden changes by single mutations concerning the rate of certain embryological processes."

As has been evident in each chapter of this monograph, the scientific belief in the fact of evolution is necessarily based on a network of a number of converging lines of evidence. The evidence is nonetheless indirect—no one has yet succeeded in bringing into existence a new species, either by macromutation or neo-Darwinian micromutation. Fruitfly (*Drosophila*) mutants have been selectively bred for thousands upon thousands of generations without producing anything more than a new strain or variety of *Drosophila,* thereby unintentionally supporting Bateson's and Goldschmidt's view that subspecies are not incipient species (in contrast to Darwin's hypothesis in 1859). Thus, while even the sternest critics of neo-Darwinism believe that "the major facts and explanations of evolution by mutation, selection, and adaptation are established forever" (Goldschmidt, 1952), the problem seems remarkably resistant to experimental verification.

◇ ◇ 10 ◇ ◇

R.A. Fisher, J.B.S. Haldane, and Sewall Wright: The Genetics of Populations [1]

R. A. Fisher (1890–1962): The Merger of Mendelian Discontinuous Heredity with Pearsonian Continuous Variation

In 1918, R. A. Fisher published an article on "The Correlation between Relatives on the Supposition of Mendelian Inheritance," in which he made a theoretical synthesis of Karl Pearson's biometry with Gregor Mendel's factorial or particulate inheritance. The greatest stumbling block to the acceptance of natural selection as the basis for evolution, other than its not dealing with the origin of variations, was Darwin's own contention that inheritance between extreme types of parent tended to blend so that variation outside of the norm of the population could not be retained from generation to generation. Mendel's discovery of so-called particulate inheritance in peas offered a means by which recessive genes could be carried along undiluted from generation to generation until they met with another recessive and then the otherwise hidden genetic potential would express itself in a phenotypic trait or characteristic. In 1902,

Yule had shown mathematically that Mendelian variation could be retained indefinitely in a randomly breeding population. It was on this basis that Fisher was able to synthesize the idea of continuous phenotypic variation of the biometricians with the genetic discontinuity of the Mendelian mechanism of inheritance. By devising specific statistical procedures that went beyond those of Karl Pearson for measuring correlation and variance among relatives, Fisher was able to show theoretically that the genes were much more influential than the environment in, for example, determining stature in siblings. This erroneous thinking is, of course, a logical extension of Galton's nature–nurture dichotomy. In fact, Fisher came to the amazing conclusion that "it is very unlikely that so much as 5 per cent of the total variance is due to causes not heritable" (Fisher, 1918). He was able to come to this conclusion by assuming that many genes contributed to a continuously varying phenotype such as stature. (Although Fisher was no doubt correct in his assumption that many genes contribute to a complex physical feature such as height, as will be documented in a subsequent chapter on the interactional and supragenetic nature of individual development, the amount of variation potentially available in this physical feature, even in identical twins, renders dubious Fisher's notion that differences in stature are 95+ percent determined by genetic variance.)

Later on, Fisher (1930/1958), was able to show statistically, first, that if natural selection favored a relatively rare gene for a long enough period, that gene would eventually become very widespread in the population. This was the perfect complement to Darwin's notion of the gradualness of evolution by natural selection. The importance of single-gene substitutions in polygenically determined traits is that they would eventually change the polygenic trait and thus could, in theory at least, be responsible for gradual evolutionary change. Single-gene thinking was "in the air" around the 1920s for other reasons, particularly because the experimental work of Thomas Hunt Morgan's group of geneticists and embryologists was guided by the assumption that evolution proceeded by the replacement of one gene by another ("point mutation") in a population of interbreeding organisms. Several other mathematically sophisticated scientists were working to show by statistical means how dominance could (theoretically) evolve depending on the selection regimen.

Fig. 10–1. Ronald A. Fisher (1890–1962).

Sewall Wright (1889–1988): Polygenic Systems, Genetic Drift, Pleiotropy

Among the most important empirical contributions on the manifestation of genetic variability in the face of strong selection (i.e., selective breeding) were the results of inbreeding (brother–sister

Fig. 10–2. Sewall Wright (1889–1988).

mating) experiments with guinea pigs reported by Sewall Wright (1920, 1922). While inbreeding automatically leads to less genetic variation (homozygosity) and reduced vigor or viability in the inbred family, each inbred family presented Wright with a somewhat different but stable constellation of phenotypic characteristics (coat-color pattern, number of digits, vigor, particular abnormalities,

Fig. 10–3. J. B. S. Haldane (1892–1964).

etc.). Consequently, it could be said that in each family different complexes or combinations of genetic factors were fixed by selection over the course of the twenty plus generations of his experiment. Since this was the case, Wright became an advocate of the notion that each phenotype was affected by multiple genes (polygenic systems), and he advanced the even more novel idea that the same genes could affect more than one phenotype depending on the genetic network or genetic interaction system in which they found themselves (pleiotropic effects). The significance of Wright's empirical results for population genetics and evolution is that (1) each of the inbred families could be seen as small interbreeding subpopulations, that (2) selection (in this case, inbreeding of brothers and sisters) brought about different phenotypic end results in each subpopulation, and, thus, (3) each subpopulation ended up being genetically different even though the genetic starting point was the same. The fact that each subpopulation "drifted" in its own nonpredictable way (i.e., there was a chance fixation of different combinations of genes in each family group) eventually gave rise to Wright's idea of "genetic drift" in small populations, which could offer natural selection the kind of systemic genetic variation that would provide the basis for a fairly rapid evolution (Wright, 1980). This was in contrast to Fisher's idea that evolution proceeded slowly by single-gene substitutions in large interbreeding populations. (Of course both points of view could be valid.)

J. B. S. Haldane (1892–1964): Single-Gene Mutants

By way of rounding out the various schemes and assumptions of the new genetics of populations, J. B. S. Haldane (1924; 1932a) supplied a statistical analysis of the consequence of the introduction of favorable single-gene mutants or "sports" into an interbreeding group in which the single gene was responsible for a new trait. Neither Fisher nor Wright had considered these kinds of single-gene mutations in their approaches, but such mutations are obviously quite compatible with a simple Mendelian scheme of particulate inheritance and evolution.

In summary, as shown in Figure 10–4, there are three different ways in which the founders of population genetics pictured the relationship between genes and traits: In the simple Mendelism of Haldane's single-gene mutants, one gene underlies one phenotype. In Fisher's polygenic or multifactorial theory, a number of genes contributes to a single trait. Finally, in Wright's theory, each trait is affected not only by a number of genes but each gene affects more than one trait (pleiotropy), and genic expression is influenced by the

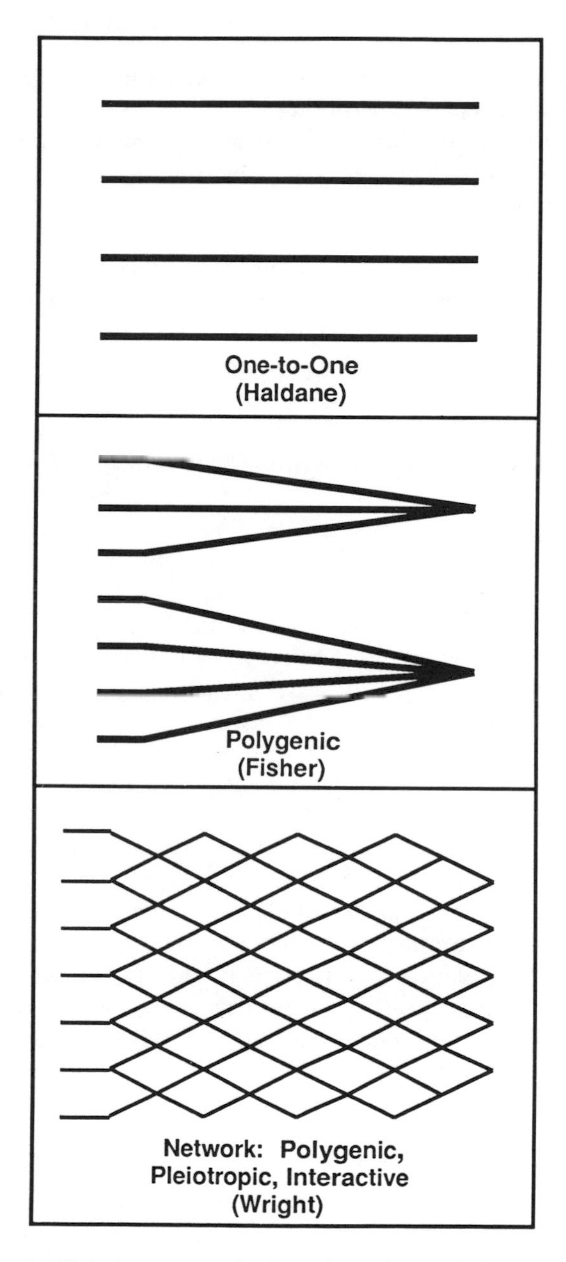

Fig. 10–4. The three ways the founders of population genetics saw the relationship between genes and phenotypic traits. (Adapted from Wright, 1980.)

genetic network that the genes find themselves in (i.e., genetic interaction itself affects gene expression).

The various approaches to population genetics reviewed here could be viewed as complementary, rather than mutually exclusive, in the sense that each author is addressing or emphasizing a different aspects of population genetics. In fact, there are internal inconsistencies that have not been resolved to this day (Provine, 1971), but I think it is more pertinent in the present context to ask about the role, if any, of individual development that was allowed for in the largely statistical conceptions of population genetics.

Getting from Gene to Organism in Population Genetics

As described previously, early in the 1900s there was not only an experimental but a conceptual separation of the study of genetics from the study of embryology, so that truly "getting from gene to organism" would be (and still is) precluded in all but the most abstract ways. Among the three founders of population genetics, as far as I can determine, Fisher devoted virtually no public thought to the problem, while Haldane approached the issue briefly in two different ways, and Wright lived long enough into the twentieth century (he was still alive and actively publishing scientific articles in 1984) to give a fairly differentiated view of what would be required to synthesize gene and phenotype. It is fair to say that the developmental pathway between gene and organism has not ever been a major concern for most population thinkers. It is not merely that it is a different problem, but that it is seen as peripheral to the main aims of population genetics, which are (1) to work out a formal or mathematical theory of the gene as the basis of transmission genetics for within-species intergenerational correlations (heritability or h^2) and (2) to calculate the hypothetical changes in gene frequencies in populations within a species that would be required for evolution to occur. Since natural selection operates on individuals and particularly on individual phenotypes, and individual phenotypes are a consequence of ontogenetic development, a concern for individual development is highly relevant to population genetics and evolution even if it is given short shrift by most population theorists. That is to say, natural selection acts as a consequence of individual development, so it cannot provide the explanation or cause of individual development, as is sometimes implied by population-genetic thinking.

Although Haldane (1946) did not explicitly (formally) recognize a developmental factor, in his purely logical formulation of the four principal types of interaction between nature and nurture he implicitly invokes such a factor. He begins his logical explication by supposing there are two genetically different populations *A* and *B*, and two different environments *X* and *Y*. If we do experiments in which *A* and *B* are reared in environments *X* and *Y* and we measure any phenotypic outcome of such an experiment (e.g., number of eggs laid, milk yield, color intensity, dominance in combat, or intelligence), assuming a "best" can be assigned to these phenotypic outcomes, we will get the following possibilities:

Type 1*a*		*X*	*Y*	Type 1*b*		*X*	*Y*
	A	1	2		*A*	1	3
	B	3	4		*B*	2	4
Type 2							
	A	1	4				
	B	2	3				
Type 3							
	A	1	2				
	B	4	3				
Type 4*a*				Type 4*b*			
	A	1	3		*A*	1	4
	B	4	2		*B*	3	2

Thus, arrangement 1*a* signifies that strain or subspecies *A* in environment *X* did best (e.g., produced the most eggs or milk, had the highest measurable intelligence), while strain *A* in environment *Y* did second best and so on. For simplicity, the above assumes that *AX* is the best combination, so as to give only six outcomes rather than a near infinity of possible outcomes.

Haldane used this format to demonstrate that there are at least six possible types of nature–nurture interaction even under the unrealistically simplest situation (*AX* always best with only two pure genotypes and two different environments). Haldane was able to find examples of each type of interaction from the human eugenics and agricultural breeding literatures. *His main point was that given the information on the outcomes in this simplest case, it was not really possible to predict what would happen under any other environmental condition,* nor even to advocate long-term breeding practices even if the tangible environment were held constant, other than to make the logical argument that it would be wise to maintain a high level of genetic diversity in the event of an environmental change.

Without belaboring the point, it goes almost without saying that it is only when one knows considerable about the influence of specific developmental factors as well as the tangible environment that some measure of prediction and control are possible even in the agricultural area and certainly with respect to evolution. Earlier, Haldane (1932b), following de Beer, had shown some interest in development in the sense of seeing how heterochrony could provide an impetus to evolution. He, thus, speculated on the significance of the time of action of different genes and discussed how selection of accelerating and decelerating temporal alterations in genetic activity could have significant evolutionary implications depending on when they occurred in the entire life cycle (e.g., the earlier the more profound the effect). Goldschmidt (1938) and Ford and Huxley (1927), among others, also viewed the genes as determining rates of various physiological and morphological events during individual development. But beginning with the neo-Darwinian synthesis in the 1940s, this point of view has been underrepresented in favor of viewing the genes as establishing a structural blueprint for development and as setting strict limitations or constraints on individual development. Wright's more dynamic physiological conception of the gene (next section) has not yet taken hold even today. The theoretical treatment of genes in population genetics has so won the day that evolution is now defined in terms of changes in gene frequencies in populations, even though these changes are entirely hypothetical!

Wright on Heredity, Environment, and Development

It is most interesting that Galton's concepts of nature and nurture survive to the present day as the principal terms of contemporary population genetics. As we follow the concepts of nature and nurture as they become formalized in Pearson's biometry and Fisher's genetic theory of natural selection and beyond, at the present time we have the notions that variations in genes or heredity (h^2) and variations in the environment (e^2) are taken to be the two exclusive sources of observed phenotypic variations in populations. In light of this historical development, the significance of which cannot be overestimated, it is fascinating that Sewall Wright, in one of his first papers on the subject, found that he needed not only heredity (h^2) and what he called the tangible environment (e^2) but a developmental factor (d^2) to understand the course of coat-color differences in the stocks of guinea pigs he studied for over twenty generations (Wright, 1920). Since despite intensive inbreeding no particular patterns of coat color ever became well fixed—that is, variations covering almost the entire range from solid color to solid white were continuously found within each inbred line—Wright was forced to conclude that a dominant-gene factor was lacking with respect to the determination of coat color. It occurred to him that there must be irregularities of individual coat-color development not under strict genetic control so he factored in "irregularity of development" to attempt to account for the results of his selective breeding experiment. When he did the calculations, Wright found that in the control (randomly bred) stock coat-color variations were determined 42 percent by heredity and 58 percent by irregularity of development, leaving no variation ascribable to tangible environmental factors (feed, weather, health of dam, etc.). The developmental factor was even stronger in the inbred line, where only 3 percent of the variation was ascribable to differences in heredity (they were almost completely homozygous), 5 percent to tangible environmental differences, and 92 percent to irregularities in individual development. It is most unfortunate that Wright's tripartite scheme did not take hold, as it could have encouraged a way of thinking that would have possibly overcome the inadequacies of the more simple view that the phenotype or phenotypic differences are a consequence of only genetic and tangible environmental causes. The genuine conceptual significance of Wright's need to incorporate developmental considerations into his genetic framework for thinking about heredity became even clearer after the discovery of DNA and the identification of the genic material as DNA. Now, at least in Wright's system, DNA (i.e., the genes) explicitly became part of the total developmental–physiological system for producing an organism, entailing all the internal and external interactions alluded to in the chapter

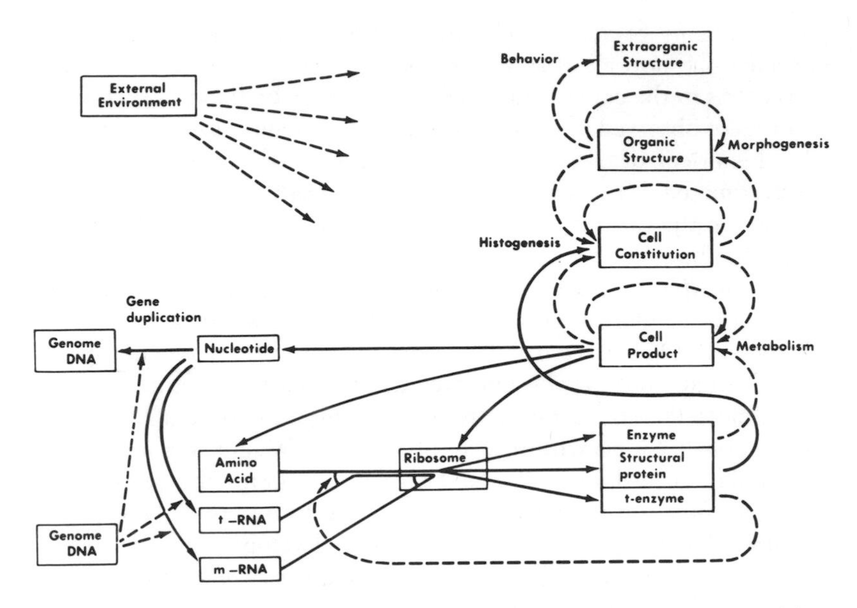

Fig. 10–5. The fully interactive developmental system, as presented by Wright (1968). Wright appears to be the only geneticist of his era to have held a truly developmental viewpoint.

that discussed the developmental considerations of de Beer. Figure 10–5 is the way Wright (1968) depicted the developmental–physiological system as it includes the genic material.

Without going into exhaustive detail on all the events in Figure 10–5, for the present purposes it is useful to observe that the genes are not given a status outside of the interacting system of development and thus the activity of the genes (DNA) themselves are subject to feedback effects during the course of individual development. Here is the way Wright expressed it:

> It was formerly common for biologists . . . to attribute the development of characters partly to physiological processes and partly to heredity, as if heredity could operate by some sort of sympathetic magic, independently of the physiological channels. The attitude of physiological genetics has been that characters are *completely* determined by physiological processes [and] that the genes are . . . the ultimate internal physiological agents.

> If, instead of tracing forward from primary gene action [Figure 10–5] we consider everything that may affect a particular process at a particular time in development, these fall naturally into four categories: (1) local gene action, (2) the chain of past events in the line of cells in question, (3) correlative influences from adjacent cells and from other parts of the body, and (4) external environmental differences. Since the second and third may be analyzed, step by step, on this fourfold basis, and local gene action must be evoked by products of previous events, the ultimate factors are the array of hereditary entities in the egg and sperm and the succession of external influences. (Wright, 1968, pp. 58–59)

This is, of course, getting ahead of our story and perhaps in part explains why Wright's multifaceted theorizing has not been as influential as Fisher's and Haldane's more simple and straightforward models in population genetics.[2] Both Fisher and Haldane thought of the phenotype and phenotypic differences as being caused by what is now routinely called hereditary (h^2) and environmental (e^2) sources. Since in this simpler scheme the contribution of the two sources equals unity (100 percent), if you can estimate or calculate the percentage of one you can derive the percentage contribution of the other. Although population thinkers tell us that, strictly speaking, h^2 and e^2 refer to sources of individual *differences* among phenotypes, as a matter of fact in actual practice these measures are often applied to a causal understanding of the outcome of individual development as well.

Though it is possible to give a long list of examples of the inappropriate application of h^2 and e^2 to the causal understanding of an individual phenotype, I give only one here as an exemplar from leaders in the psychiatric genetics literature having to do with schizophrenia: "At this time we are willing to suggest that the genetic heritability of the liability for schizophrenia is about 70%, and the cultural-inheritance component accounts for about 20% of the combined liability" (Gottesman & Shields, 1982, p. 229). Since these authors are quite sophisticated behavioral geneticists of very high reputation, one may well wonder what these figures are intended to tell us about the course(s) of schizophrenia. On the next line the authors go on to say: "It is important to understand the implications of finding that a trait such as the liability to schizophrenia has a high heritability. In the samples so far studied, it means that environmental factors were relatively unimportant as causative agents of the schizophrenias. However, . . . these data do not permit the conclusion that curative or preventive measures will be ineffective" (Gottesman & Shields, p. 230). Aside from the fact that the meaning or significance of the 70 percent figure is inscrutable in terms of individual development, the main point of the quotation is to show that h^2 and e^2 are applied to individual phenotypic outcomes and not merely to phenotypic differences in populations.

The dichotomous idea of the exclusive and precise genetic or environmental determination of traits in individuals is, of course, an indubitable legacy of Galton's thinking that the outcome of individual development is determined so much by nature and so much by nurture. If h^2 actually was useful for estimating genetic constraints or limitations on developmental outcomes, it would have some value, but it is widely agreed by geneticists themselves that h^2 cannot be interpreted in that way (Feldman & Lewontin, 1975), although individual scientists may now and again lapse into thinking in those terms (e.g., Gottesman & Shields).[3] The important distinction, which is so often lost, is that developmental analysis is required to get at the factors that determine phenotypic traits and population-genetic analysis is often said to deal only with the factors that cause *differences* (variation, variance) in traits. An Oriental acquaintance has commented that the meaning of heritability (h^2) boggles the average mind. She noted that all Chinese have black hair and thus the heritability of black hair among Chinese is 0! Clearly, it is difficult for scientist and layperson alike to remember that heritability is supposed to refer to the causation of differences in traits and not to the development of the traits themselves. A further cause for confusion is that population geneticists sometimes speak as if there are correlations between genes and traits: "Essentially, h^2 is a measure of correlation between genotype and phenotype" (Lerner, 1968, p. 144). So, while population geneticists may not be confused by Professor Lerner's statement, such a remark would almost certainly be confusing to lay people and scientists in other fields. (It would have been clearer and more correct for Professor Lerner to have said h^2 is a measure of correlation between genetic variation and phenotypic variation.) On the other hand, these three sentences occur on the same page of a population geneticist's book: "Genetic influence is imbedded in the complexity of interactions among genes, physiology, and environment. It is probabilistic, not deterministic; it puts no constraints on what could be. . . . some individuals are genetically predisposed to abuse [alcohol]" (Plomin, 1986, p. 21).

As mentioned earlier, it is essential to remember that selection (whether considered as survival or selective mating or breeding) operates as a consequence of development, so it cannot be the cause of development.

◇ ◇ 11 ◇ ◇

Evolution: The Modern Synthesis and Its Failure to Incorporate Individual Development into Evolutionary Theory

T. Dobzhansky (1900–1975): Evolution Defined in Terms of the Genetics of Populations

In 1937, Theodosius Dobzhansky published a book called *Genetics and the Origin of Species,* which is widely regarded as ushering in an intellectual era in biology, hailed as the triumph of neo-Darwinism, and called the *modern evolutionary synthesis*. Dobzhansky integrated his experience as an experimental geneticist in the laboratory with the observations of naturalists working on populations in the field to show that populations of similar individuals contained sufficient genetic diversity for natural selection (differential reproduction) to theoretically bring about evolution in a gradual fashion and this without the assistance of the dubious

Fig. 11–1. Theodosius G. Dobzhansky (1900–1975).

mechanism of the inheritance of acquired characteristics. In other words, Dobzhansky, using Weismann's and Mendel's concept of "hard inheritance"—genes are unalterable by the organism's life experience and genes give rise to characters—forged a population-genetic view of evolution that overcame Darwin's problem of blending inheritance and the need to appeal to Lamarck's use-inheritance.

The best adapted organisms in a population had the most surviving offspring and thus contributed more of their genetic material to the next generation ad infinitum until, eventually, a new species would arise as a result of this long-continued process of selection of the best adapted individuals and the favoring (even slightly) of their unique genotypes into the indefinite future. A very important theoretical component of the modern synthesis is the idea that some organismic variation is heritable and some is not heritable and it is only the heritable variation that is significant for evolution. This view stems inevitably from population-genetic thinking, which holds that there are two sources of variation, one genetic and the other environmental. According to this viewpoint, it is only the former, of course, that can be inherited. (As we shall see later in this chapter, this overly simple idea requires substantial modification, but for the period we are describing it worked very well.)

In the third edition of his book, in 1951, Dobzhansky identified evolution as occurring at the level of populations and, more specifically, in the genetics of populations: "Evolution is a change in the genetic composition of populations. The study of mechanisms of evolution falls within the province of population genetics" (p. 16). The rules that govern the genetic structure of populations are different from the rules that govern the genetics of individuals.

> A Mendelian population may be said to possess a corporate genotype. The population genotype is evidently a function of the genetic constitution of the component individuals, just as the health of an individual body is a function of the soundness of its parts. The rules governing the genetic structure of a population are, nevertheless, distinct from those which govern the genetics of individuals, just as the rules of sociology are distinct from physiological ones, although the former are in the last analysis integrated systems of the latter. (Dobzhansky, 1951, p. 115)

Ernst Mayr (1904–): Genetic and Nongenetic Sources of Phenotypic Variation

Ernst Mayr, another significant architect of the modern synthesis, in his 1942 book, *Systematics and the Origin of Species,* laid great stress on the point that variation in populations—the variation among individuals—stems from two sources: nongenetic (phe-

Fig. 11–2. Ernst Mayr (1904–).

notypical) variation and hereditary (genotypical) variation. This is, of course, a fundamental tenet of population genetics, as we saw in the previous chapter describing the birth of that field and in an earlier chapter on Galton in which the nature–nurture dichotomy was first broached in a systematic way. While it is certainly correct to say that identical twins, one reared in America the other in France, came to speak, respectively, English and French because of the predominant language environment to which they were exposed, it is not correct to think that the phenotypic outcome of learning to speak French or English is nongenetic (i.e., it does necessarily have a genetic component, no matter how indirect the genetic component may seem to be). Of the three areas of genetic research, (1) developmental genetics (gene action in ontogeny), (2) Mendelian or transmission genetics (parent–offspring genotypes), and (3) population genetics (the genetic composition of groups), it is only the latter two that have been incorporated in the modern synthesis of evolutionary thought. Not only has there been comparatively little work on the tremendously knotty problem of developmental genetics, the modern synthesis holds that individual heredity is unalterably fixed ("hard," not Lamarckian "soft" inheritance) and it is populations, not individuals, that evolve. Therefore, according to this kind of thinking, developmental genetics could be finessed and the synthesis could turn on the genetics of populations.

Julian Huxley (1887–1975): The Modern Synthesis

Another architect of the neo-Darwinian triumph, one who gave it its name—Evolution: The Modern Synthesis—was Julian Huxley, who, in 1942, published a book with that very title. In an introduction to a re-edition of that work, in 1963, Huxley succinctly summarized the modern synthesis as follows:

> The existence of an elaborate self-reproducing code of genetical information ensures continuity and specificity; the intrinsic capacity for mutation provides variability; the capacity for self-reproduction ensures geometric increase and therefore a struggle for existence; the existence of genetic variability ensures differential survival of variants and therefore natural selection; and this results in evolutionary transformation.

It is telling that in Chapter 9 of his 1942 book, Huxley has a section titled "Consequential Evolution: The Consequences of Differential Development" in which he discusses J. B. S. Haldane's notion (described in Chapter 10 here) that genes control rates of development and thus selection, theoretically operative at all stages of the life cycle, could change rates of development and thereby produce evolutionary changes. Although it is a minor and small segment of his 645-page book, Huxley was one of the few architects

Fig. 11–3. Ivan Ivanovitch Schmalhausen (1884–1963).

of the modern synthesis to even mention changes in ontogeny as contributory to the new outlook on the causes of evolution.

The brilliant architects of the modern synthesis were certainly not ignorant of the role of gene action in development, it is just that population-genetic thinking made them prone to put the analysis of evolution in terms of heritability, genes, and environments rather than in developmental terms, something that remains with us today, as population-genetic thinking is still the dominant mode of contemporary biological thought concerning evolution. As long ago (or as recently!) as 1971, the geneticist C. A. Thomas wrote: "We need a [developmental] guide to understanding speciation and evolution. The hundred-year-old notions of mutation and selection (however correct they may be) no longer offer us any insights into the central issue of biology" (Thomas, 1971, p. 237).

I. I. Schmalhausen (1884–1963): Ontogenetic Adaptability and the Evolution of Adaptations

The only contributor to the modern synthesis who saw a significant role for individual ontogenetic development in evolution was the Russian biologist Ivan Ivanovich Schmalhausen. In 1949, an English translation of his book appeared in the United States, *Factors of Evolution: The Theory of Stabilizing Selection,* under the supervision of Theodosius Dobzhansky, himself a Russian expatriate. Dobzhansky, in the foreword to Schmalhausen's book, credits the latter with going beyond the synthesis achieved by Dobzhansky and the other principal architects of the triumph of neo-Darwinism (Ernst Mayr, Julian Huxley, George Gaylord Simpson, Bernhard Rensch, G. Ledyard Stebbins, Jr.).[1] Dobzhansky writes, "Schmalhausen advances the synthetic treatment of evolution starting from a broad base of comparative embryology, comparative anatomy, and the mechanics of development. It supplies, as it were, an important missing link in the modern view of evolution."

Building on the experimental and theoretical work of his foremost teacher, A. N. Severtsov, Schmalhausen's essential insight is to see evolution as a process wherein favorable *adaptabilities* instigated by the environment eventually become genetically assimilated *adaptations,* in the sense of moving from external dependency to complete internal control. Schmalhausen's idea is that organisms have a tremendous genetic reserve that is only tapped and revealed when the organism encounters a new environment (a change of climate, food, new competitors for resources, etc.). In stark contrast to other contributors to the modern synthesis, Schmalhausen sees that adaptability itself has genetic correlates, whether through mutation or otherwise, so a favored adaptability can immediately become, in a single generation, the subject of natural selection. The long-term operation of natural selection eventually leads to changes in the developmental control factors that give rise to the feature in question, shifting the locus of control from the external environment to the internal workings of the organism, an outcome now known as the "genetic assimilation of an acquired character." Here is the way Schmalhausen (1949, p. 175) introduced his discussion of what he called "elementary dependent reactions and their transformation":

> The individual adaptability of the organism creates for it a condition of what may be called flexible stability or lability. This condition enables the organism to survive sudden and considerable variations of the external environment. Also, it can thereby actively migrate from one environment to another and even reorganize its structure. Hence, a knowledge of the development of the organism's system of adaptive reactions is very important in understanding the laws of evolution. Finally, the origin of adaptability is an insufficiently studied aspect of evolutionary theory. [Why? Because] The Larmarckians based their theory upon the premise of an already existing individual adaptability and did not examine its origin while the neo-Darwinians regarded it as unimportant since they assumed that the results of individual adaptability, *being nonheritable,* have no evolutionary value. (italics added)

Shades of Mayr's two sources of variation in individuals! It is telling that, while Schmalhausen is sometimes recognized as one of the architects of the modern synthesis, his idea of the significance of individual adaptability has not been incorporated into the modern synthesis. More on this later in connection with the Baldwin Effect.

Given Schmalhausen's view that the environmentally provoked adaptabilities themselves have a genetic basis, and that it is possible eventually to shift the dependence from external to internal sources in the course of selective breeding (stabilizing selection II to be further described below), it is interesting that the necessary experi-

Fig. 11–4. James Mark Baldwin (1861–1934), *top left,* Conwy Lloyd Morgan (1852–1936), *top right,* and Henry Fairfield Osborn (1857–1935), independent coauthors of what came to be called the Baldwin Effect.

mental proof for that transformation was not to arise until four years after the publication of his book in the English language . . . and it was a British developmental geneticist, C. H. Waddington (1905–1975), who, in 1953, was to provide the compelling evidence.[2] By way of setting the stage for an enormously important experimental result that continues to the present day to lie—in a conceptually incompatible way—outside the modern synthesis, it will be useful to hark back some eighty-five years to the first intimations of what is now called the *Baldwin Effect* and also known as the *genetic assimilation of an acquired character.*[3]

The Baldwin Effect

The comparative developmental psychologist James Mark Baldwin (1861–1934) was among the few scientists preceding Schmalhausen to appreciate the evolutionary significance of individual differences in adaptability. In fact, Baldwin published a book on the topic—*Development and Evolution*—in 1902, to which Schmalhausen himself (pp. 197–198) makes a glowing, if fleeting, reference. As was well understood at the time, individual differences (variation) are at the root of the Darwinian theory of evolution. Those individuals that are most adapted leave the most offspring and thus their genotypes shape, circumscribe, and give "direction" to the capabilities of the next generation, and so on, indefinitely. What Baldwin perceived, along with the comparative psychologist Conwy Lloyd Morgan (1852–1936) and the biologist Henry Fairfield Osborn (1857–1935), was that individuals differ in their modifiability or adaptability to changed circumstances, and it is those who can adapt that differentially affect the course of the next generation's adaptability and thus shape the direction of future lineages. Here is the way Baldwin (1902) expressed it:

> The variations which have been utilized for ontogenetic accommodation in the earlier generations, being thus kept in existence, are utilized more widely in the subsequent generations. Congenital variations, on the one hand, are kept alive and made effective by their use for adjustments in the life of the individual; and, on the other hand, [this is the new notion] adaptations become congenital by further progress and refinement of variation in the same lines of function as those which their acquisition by the individual called into play. But there is no need in either case to assume the Lamarckian factor. (p. 98)

The role of intuition in science is no better in evidence anywhere than in the assertion, to paraphrase, "ontogenetic adapta-

Fig. 11–5. Conrad H. Waddington (1905–1975), who performed the brilliant experiments demonstrating the so-called genetic assimilation of an acquired character.

tions become congenital without the Lamarckian factor." Baldwin adduces no indirect, much less direct, evidence for this view, and half a century was to go by before Schmalhausen made the idea the cornerstone of his stabilizing selection and C. H. Waddington was to demonstrate it experimentally.

In his now-famous experiment, Waddington (1953) subjected fruitflies to a heat shock during their embryonic (pupal) development. Some of the surviving flies responded to the brief high-temperature shock by developing wings with few or no crossveins in them. (This has no known adaptive value; it was merely a convenient response for Waddington to work with.) He then selected these particular phenotypes for breeding and subjected their progeny to the heat shock, once again selecting the crossveinless phenotypes for breeding. He repeated this procedure for fourteen generations, by which time a small proportion of the descendant fruitflies so selected exhibited the crossveinless phenotype in advance of the exposure to the heat shock—just the result that is necessitated by Baldwin's hypothesis and Schmalhausen's more comprehensive theory of stabilizing selection II (described further on page 133).

It is important to note that Waddington's result has all the earmarks of the inheritance of an acquired character—were we unaware that selective breeding had been exercised on the fruitflies, we would have had to assume the operation of the nineteenth century view of the inheritance of acquired characteristics. The amazing outcome is the development of crossveinlessness in the fourteenth generation without the presence of the environmental stimulus that originally gave rise to the condition in the founding and subsequent generations. Had Darwin suspected the possibility of such an outcome by natural selection, he would not have had to retain the untenable Lamarckian "mechanism" alongside natural selection as a mode of evolution.[4]

The Baldwin Effect, as the Baldwin–Morgan–Osborn hypothesis came to be called, has floated in a curious conceptual limbo to the present day in biology. I think it is clear why it has not become prominently and regularly incorporated into discussions of evolutionary mechanisms, anymore than that same feature of Schmalhausen's theory: some leading biologists have the idea that adaptability or plasticity (modifiability) is not genetically based or correlated. For example, Ernst Mayr (1958), in a chapter in a book on behavior and evolution, defines the Baldwin Effect as "the hypothesis that a nongenetic plasticity of the phenotype facilitates reconstruction of the genotype" (p. 354). Elsewhere in the same volume the Baldwin Effect is defined by another author as "the hypothesis that nongenetic adaptive modification of the phenotype may be replaced by genetically controlled modification" (Emerson,

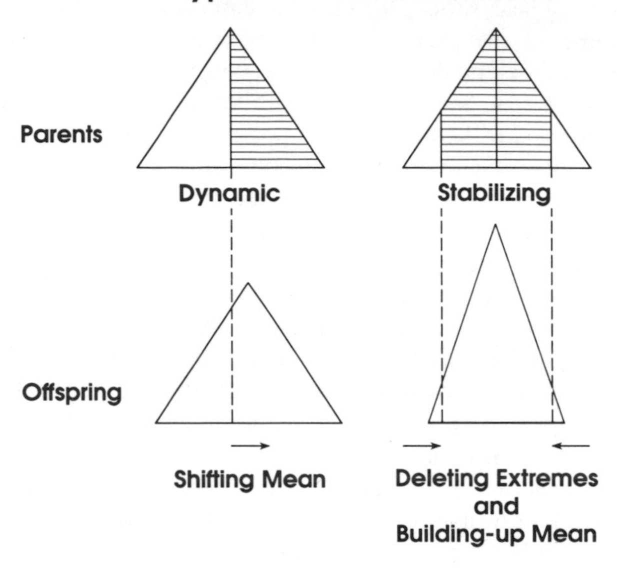

Fig. 11–6. Schematic depiction of dynamic and stabilizing forms of natural selection. Dynamic selection presents the nub of Darwin's idea of what natural selection accomplishes, whereas stabilizing selection represents Schmalhausen's contribution to population-genetic ideas undergirding the modern synthesis. Schmalhausen also used the term stabilizing selection for quite another phenomenon, one akin to the Baldwin Effect or genetic assimilation. The second usage of stabilizing selection has not yet found its way into the modern synthesis.

1958, p. 318), even though the latter author perceives a relationship to Waddington's work. The idea that behavior or any other organismic manifestation can be "nongenetic" in any sense of the word is incorrect.[5] Waddington himself failed to appreciate the conceptual linkage between the Baldwin Effect and his genetic assimilation—in fact, he saw the revival of Baldwin's idea by Simpson (1953) as a symptom of misguided American nationalism (Waddington, 1969, p. 386; Waddington, 1975, p. 88)!

Essential Similarity of the Baldwin Effect, Stabilizing Selection, and Genetic Assimilation

Schmalhausen's book contained two versions of stabilizing selection, one of which has been regularly incorporated into the population-thinking of the modern synthesis and the other of which has languished on the same conceptual vine as the Baldwin Effect and genetic assimilation as far as evolutionary theory is concerned.

The meaning of stabilizing selection that all evolutionary biologists are quite familiar with is shown in Figure 11–6. Schmalhausen discriminated between a "dynamic form" of natural selection (the familiar Darwinian idea that some variants are favored over others as far as leaving progeny is concerned) and the stabilizing form of natural selection that eliminates the most extreme forms of variation and builds up the mean or average form by selecting against the extremes at both ends of the distribution. One can picture this process as dynamic or directional selection favoring a segment of the normally distributed population in the face of environmental change (left side of Figure 11–6) and stabilizing selection coming into play in subsequent generations to maintain the variation around the new mean (right side of Figure 11–6). These sorts of selection operating on populations are the very stuff of the modern synthesis, so what I shall call Schmalhausen's first meaning of stabilizing selection (SS I) has found a well-recognized place in conventional evolutionary biological thinking.

Schmalhausen's second meaning of the term stabilizing selection (SS II)—the one mentioned earlier in this chapter—has not become part of the standard evolutionary theory for the very same reason that the Baldwin Effect and genetic assimilation have not been incorporated into the modern synthesis: phenotypes produced by the environment are erroneously seen as nongenetic and thus have no place in the modern synthesis.

The essential similarity of the Baldwin Effect, SS II, and genetic assimilation is the notion that (1) a severe change in the environment exerts a fairly drastic selective effect leaving only a fragment of the original population in place to reproduce, i.e., those having sufficient adaptability to survive; (2) as the progeny of the survivors selectively breed among themselves for some number of generations in the new environment, (3) eventually the original behavioral or morphological change produced by the new environment becomes genetically assimilated (developmentally canalized) so the trait now appears even if these organisms move to a different environment, or the stressor is removed (such as heat shock in Waddington's crossveinless experiment described earlier).

Matsuda has also recognized the essential similarity among the Baldwin Effect, genetic assimilation, and the second meaning of Schmalhausen's stabilizing selection (Matsuda, 1987, pp. 43–46). He comments on it favorably, labels it a neo-Larmarckian scenario (which places it outside the modern synthesis), and takes it as an accurate description of the way animals evolve in changing environments.

Attempt to Place an Evolutionary Change in Developmental Mediation within Mildly Expanded Tenets of Modern Synthesis

In my opinion, the above three stages are common to the Baldwin Effect, genetic assimilation (developmental canalization), and the second meaning of Schmalhausen's stabilizing selection. In order to show how this evolutionary change in the developmental mediation of a new response could be made to fit within the (mildly expanded) tenets of the modern synthesis, I shall describe one of Waddington's experiments that is more amenable to the natural selection framework of the modern synthesis than the crossveinless experiment, because it deals with an adaptive response rather than a maladaptive one. (Flies with wings lacking crossveins are very likely at a strong selective disadvantage for mating, especially since wingbeating is part of the courting behavior of fruitflies.)[6]

In another experiment, Waddington (1959, 1961) exposed developing fruitflies to food containing small to large amounts of extra salt (sodium chloride). Those flies that survived after twenty-one generations showed a large increase in the size of their anal papillae. The latter can be viewed as an adaptation, because the enlarged papillae have to do with the surviving flies' ability to handle the extra salt. Exposure to the extra salt during development caused considerable mortality (>60 percent). And the survivors bred among themselves. So, this experiment has the necessary hallmarks of the neo-Darwinian scenario for evolution: differential

Evolutionary Progression of Stages Common to Baldwin Effect, Genetic Assimilation (Developmental Canalization), and Second Meaning of Stabilizing Selection

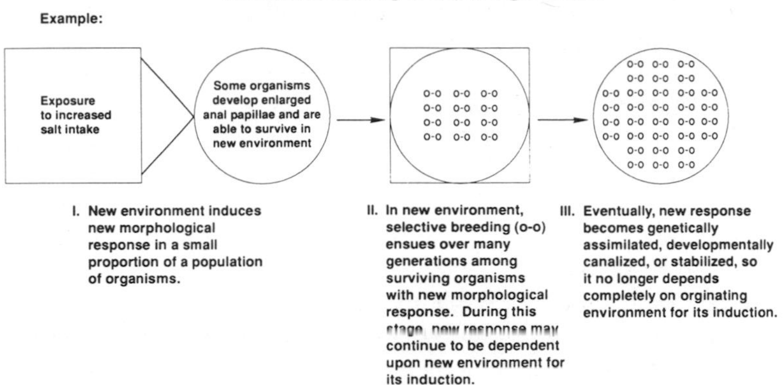

Fig. 11–7. Three stages in the developmental assimilation or autoregulation of an originally environmentally induced trait. There is a change in the developmental mediation of the trait in the third stage.

mortality and reproduction, and selective breeding among the surviving individuals showing the adaptation.

Taking some license, in Figure 11–7 I have placed the progress and results of Waddington's sodium chloride experiment within the three-stage framework previously described in connection with the commonalities among the Baldwin Effect, genetic assimilation (developmental canalization), and the second meaning of stabilizing selection. The first two stages (I and II in Figure 11–7) are not controversial from any point of view. The survivors breed among themselves, their progeny gradually increase in number over the course of generations despite continuing to be stressed by the salt environment, and the adaptation recurs from generation to generation, very likely with some improvement in the population at large in the ability to handle the extra salt. It is the third stage (III in Figure 11–7) that is unexpected and not predicted by the straightforward action of natural selection, which is to say that if the salt environment were greatly reduced or otherwise altered, the progeny would continue to show an enlargement of their anal papillae (for a period of some generations) as compared to the original stock or control flies not subjected to the experimental regimen. What Waddington's actual results do show is that at every concentration from weakest to strongest, the selected flies develop larger papillae than the unselected, control populations, testifying to some degree of genetic assimilation or developmental canalization of the originally environmentally induced trait.

It seems to the writer that it is essential to extend the tenets of the modern synthesis to include the second meaning of stabilizing selection. It is already well accepted by the concept of pleiotropy that selection almost always produces unanticipated changes in other areas that are somehow or other connected with the specific trait under selection. It does not seem so farfetched that such selection could eventuate in a change in developmental mediation of the selected trait, either through pleiotropic accompaniments or the modern synthesis's old standby: a favorable mutation. In any event, since the change in developmental mediation is an empirical fact that has been experimentally demonstrated by researchers other than Waddington (see his 1961 review), the basic tenets of the modern synthesis need to be changed to accommodate the phenomenon (cf. Matsuda, 1987, Chapter 5). In this regard, it is significant to note that genetic assimilation is a pivotal feature of Pritchard's (1986) avowedly "epigenetic theory of evolution."

In the next three chapters I shall describe the tenets of a developmental theory of evolutionary change that is lacking in the modern synthesis.

◇ ◇ 12 ◇ ◇

Extending the Modern Synthesis: Preliminaries to a Developmental Theory of the Phenotype (Phenogenesis)[1]

The emerging theme of the present work is that an understanding of heredity and individual development will allow not only a clear picture of how an adult animal is formed but that such an understanding is indispensable for an appreciation of the processes of evolution as well. This is a tall order, of course, and the writer would be glad if the present sketch provides what many practicing scientists would agree to be an appropriate abstract or conceptual orientation to the problem. The cardinal assumption of the present work is that the persistence of the nature–nurture dichotomy reflects an inadequate understanding of the relations among heredity, development, and evolution, or, more specifically, the relationship of genetics to embryology. This is merely another way of saying that to the extent that it is correct to say that alterations of ontogeny are responsible for evolution (phylogeny), an understanding of the relationship of genetics and embryology will go a long way toward clarifying the process of evolution. At the heart of this difficult synthesis is the fact noted previously that early in this century genetics and embryology were practiced as independent sciences even though both of them took for their essential datum the phenotype that was realized as an outcome of a particular pathway of development. Geneticists inferred the structure and function of genes or genetic activity based on particular phenotypic outcomes of their experiments. The experimental embryologists likewise inferred cellular and tissue interactions based upon the presence or absence of distortions produced in the normal or species-typical phenotype caused by their experimental manipulations. In all of this the gene or genetic material took on a separate existence that made it stand somewhat outside of, or distinct from, the developmental process as such. This was a very subtle intellectual event, and even today it is hard to portray it in a way that makes it readily comprehensible as an intelligible state of affairs. I know this firsthand, not only from my own personal difficulties of grappling with the synthesis of heredity and development, but from sustained interactions with colleagues whose appreciation of issues in developmental analysis I had profited from myself. One of these colleagues once told me that genetics and embryology had been conceptually integrated for a long time as could be seen, for example, in Thomas Hunt Morgan's book, *Embryology and Genetics,* published in 1934. I thought it was some obtuseness on my part that made it impossible for me to appreciate Morgan's supposed conceptual integration, so I was relieved when I read in another expert's book that the only "integration" of the two topics was achieved in the title of Morgan's book (Dunn, 1965, p. 190)! The difficulty of integrating the topics of genetics and embryology was, of course, appreciated by Morgan himself, and, as Dunn points out, Morgan was insightful as to the conceptual obstacle and a possible escape from it:

> As I have already pointed out, there is an interesting problem concerning the possible interaction between the chromatin of the cells and the protoplasm during development. The visible differentiation of the embryonic cells takes place in the protoplasm. The most common genetic assumption is that the genes remain the same throughout this time. It is, however, conceivable that the genes also are

building up more and more, or are changing in some way, as development proceeds in response to that part of the protoplasm in which they come to lie, and that these changes have a reciprocal influence on the protoplasm. It may be objected that this view is incompatible with the evidence that by changing the location of cells, as in grafting experiments and in regeneration, the cells may come to differentiate in another direction. But the objection is not so serious as it may appear if the basic constitution of the gene remains always the same, the postulated additions or changes in the genes being of the same order as those that take place in the protoplasm. If the latter can change its differentiation in a new environment without losing its fundamental properties, why may not the genes also? This question is clearly beyond the range of present evidence, but as a possibility it need not be rejected. The answer, for or against such an assumption, will have to wait until evidence can be obtained from experimental investigation. (Morgan, 1934, p. 234)

What Morgan is proposing here, in the final paragraph of his 1934 book, is that the integration of genetics with development could be achieved only when the gene is actively incorporated into the developmental process and that its activity is seen as reciprocally altered thereby (i.e., not only feedforward effects [gene → protein] but feedback effects as well [gene ← protein]). To those who viewed (and view) the integrity of the gene as absolutely constant, operating essentially outside the reciprocally interactive developmental system, the notion that "the genes are changing in some way" was, and is, indeed a radical suggestion. But if there is to be a truly developmental genetics, genes will have to be viewed in some sense as Morgan suggested (and as the geneticist Sewall Wright suggested) and thus become part of the entire system of mutual interactions that is the hallmark of embryological analysis and that characterizes the epigenetic development of the individual. Viewing the genes in this way continues to be a conceptual obstacle even for certain of the eminent biologists of today, so one would certainly not expect a ready appreciation of such a momentous insight among scientists outside the field of biology (e.g., in psychology, anthropology, or sociology). For example, on the concluding page of his opus magnum on *The Growth of Biological Thought*, the eminent evolutionary biologists Ernst Mayr has this to say:

The pathway from the DNA of the genome to the proteins of the cytoplasm (transcription and translation) is strictly a one-way track.

> The proteins of the body cannot induce any changes in the DNA. (Mayr, 1982, p. 828)

How different is Mayr's view, stated elsewhere in the same book, "that the DNA of the genotype does not itself enter into the developmental pathway but simply serves as a set of instructions" (Mayr, 1982, p. 824) from Sewall Wright's physiological view of DNA (see Figure 10–5, Chapter 10) and the view advocated here. That there seems to be some kind of physiological pathway back to the genes is suggested by the observation that genetic mutations can be induced by environmental causes such as extreme temperature or exposure to ionizing radiation. Also, the currently held notion that during individual development genes are "turned on and off" in their activity suggests again some mechanism of feedback from the products that genes produce back to the genes themselves. For example, DNA-binding proteins are known to regulate gene expression (protein → DNA). In sum, there has to be some means for the activity of genes to be regulated by events and processes occurring at other levels of the developmental pathway, otherwise it would not be possible to cause gene mutations by environmental agents, nor to "turn genes on and off" during normal individual development (i.e., as occurs in the construction of a normal individual of the species as opposed to a mutant). As an example of genes being turned on during development, in a leading textbook, *Comparative Embryology*, by Gorbman, Dickhoff, Vigna, Clark, and Ralph (1983), the schematic diagram shown in Figure 12–1 is presented to show how a steroid hormone diffuses into the nucleus of a cell to activate DNA transcription that results in protein secretion.

Thus, biologists accept the view that events originating outside the cell can activate DNA in the nucleus of the cell and get genes to express themselves as they usually do under normal conditions of development. This is no longer a conceptual obstacle. What continues to be a serious problem for developmental accounts of evolutionary change is how new or previously inactive genes come to be expressed. That is very possibly the most important or significant unknown in developmental genetics. I will now describe how normal (and abnormal) genetic expression can be influenced by events outside of the nucleus of the cell. From the standpoint of relating development to evolution, thc first step would be to get previously inactive genes to express themselves (i.e., abnormal gene expression)

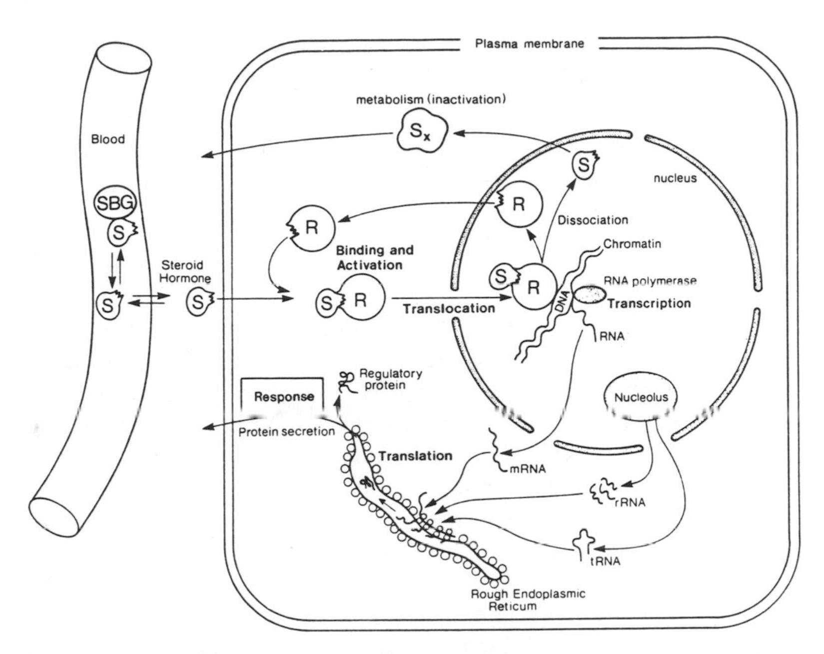

Fig. 12–1. Schematic diagram of the mechanism of action of steroid hormones effecting gene action (DNA transcription) inside the nucleus of a cell. Steroid hormone (*S*), which has dissociated from steroid-binding globulins (*SBG*) in the blood, diffuses into cell. The steroid binds to a receptor (*R*), resulting in activation and transport of the complex into the nucleus, where it activates DNA transcription. (More recent depictions hold that (*R*) is in the nucleus rather than in the cytoplasm, so now the hormone is believed to move right through the cytoplasm to the receptor in the nucleus.) Subsequently, the steroid dissociates from the receptor, leaves the nucleus, and is metabolized to an inactive form (*Sx*). (Modified with permission from Gorbman, Dickhoff, Vigna, Clark, & Ralph, 1983.)

and to have this change continue to manifest itself in subsequent generations. Regarding the question of the transgenerational stability of new phenotypes, it is essential to recall that the organism has to be constructed anew in each generation. Consequently, to the extent that the same ontogenetic factors prevail that gave rise to the new phenotype in previous generations, the changed phenotype will continue to recur in subsequent generations. Since genes do not in and of themselves make phenotypes, the same ontogenetic contingency holds for the transgenerational preservation of phenotypic changes ascribable to genetic changes.

Influence of Cytoplasm on Genetic Expression

As shown in Figure 12–1, RNA is found in the cytoplasm of the cell and both DNA and RNA are found in the nucleus of the cell, so the accepted contemporary view is that the action of DNA somehow influences RNA to produce protein by interactions in the cytoplasm of the cell. The first part of the process (DNA → RNA) is called *transcription*, and the second part (RNA → protein) is called *translation*. It is not yet understood how protein becomes differentiated into the different tissues that make up an organism during the development of an individual, but it is obvious that other interactions have to occur *above the level of the genes* to explain such differentiation (for early examples relevant to behavior, see Hydén & Egyházi, 1962, 1964), for, as stated in an earlier chapter, all the cells of the body contain the same DNA. Thus, gene expression must necessarily be different in different cells and tissues, so that different proteins are produced in the cells and tissues of the different organ systems of the body.

Based on these considerations, the most obvious candidate for influencing genetic expression during individual development is the cytoplasm of the cell, because it serves as the immediate environment of the DNA that resides in the nucleus of each cell.

Before reviewing evidence for the cytoplasm's influence on genetic expression during ontogeny, it is necessary to stand back and take a larger view of the place of the genes in the construction of an organism during individual development. As shown in Figure 12–2, the *gene* (DNA) is the ultimately reduced unit in the ever-expanding developmental pathway that moves from gene to *chromosome*—where genes can influence each other (as in Figure 10–5 in Chapter 10)—from cell *nucleus* to cell *cytoplasm*, from cell to *tissue* (organized arrangements of cells that form organ systems—the nervous system, circulatory system, musculoskeletal system, etc.), which make up the developing *organism* that interacts with other organism and, most generally speaking, the external *environment*.

The entire scheme represents a hierarchically organized system of increasing size, differentiation, and complexity, in which each

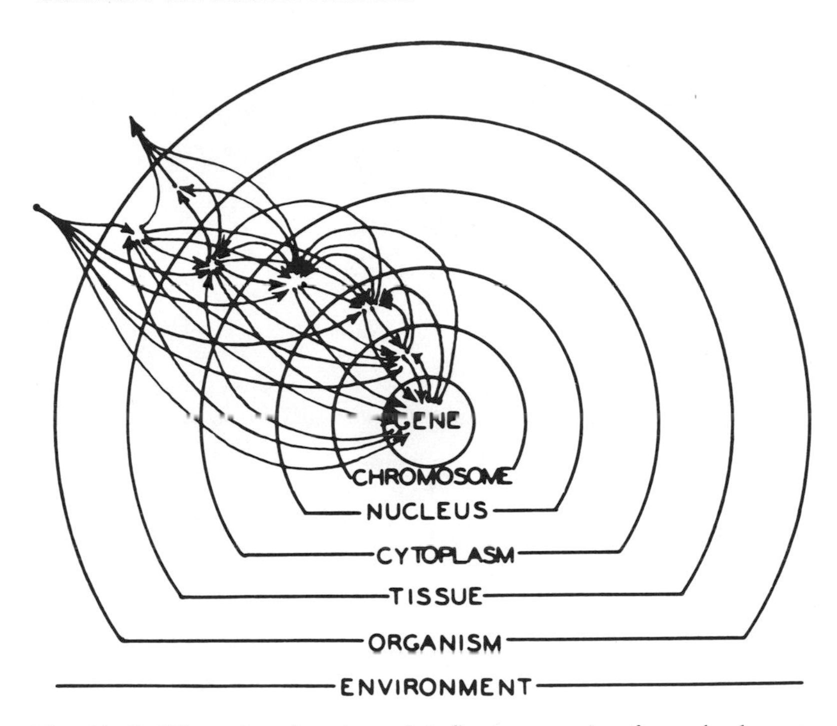

Fig. 12–2. Hierarchy of reciprocal influences moving from the lowest order of organization (gene) to the highest order (external environment). This is a somewhat less detailed depiction of the factors involved in individual development than shown in Sewall Wright's version in Chapter 10 (Figure 10–5). (From Weiss, 1959).

component affects, and is affected by, all the other components, not only at its own level but at lower and higher levels as well. Thus, the arrows in Figure 12–2 not only go upward from the gene, eventually reaching all the way to the external environment through the activities of the organism, but the arrows of influence return from the external environment through the various levels of the organism back to the genes. While the feedforward or feedupward nature of the genes has always been appreciated from the time of Weismann and Mendel on, the feedbackward or feeddownward influences have usually been thought to stop at the level of the cell membrane. The newer conception is one of a totally interrelated, fully coactional system in which the activity of the genes themselves

Fig. 12–3. Bithorax fruitfly (below) has a second set of wings that develop in place of halteres (organs of balance). (Photographs courtesy of Professor E. B. Lewis.)

can be affected through the cytoplasm of the cell by events originating at any other level in the system, including the external environment. It is known, for example, that external environmental factors such as social interactions, changing day length, and so on can cause hormones to be secreted (review by Cheng, 1979). So, in line with the diagram in Figure 12–1, such environmental events can result in the activation of DNA transcription inside the nucleus of the cell (i.e., "turning genes on").

To return to the influence of cytoplasm on gene expression, in addition to affecting gene expression in the individual, alterations of cytoplasm can have transgenerational consequences and, thus, have potential significance for evolutionary change without the alteration of genes or gene frequencies. The most recent experimental evidence for cytoplasmic influence on gene expression comes from the research of Mae-Wan Ho and collaborators (1983), in which fruitfly embryos were exposed to ether during a certain period in their early development, an experimental treatment known to result in a somatic mutation—an extra set of wings—called the bithorax phenotype (shown in Figure 12–3).[2] To rule out genetic selection for the bithorax response, the treated individuals were bred at random (i.e., individuals were allowed to mate whether they showed the bithorax phenotype or not). Across the six generations of the experiment, in which each generation was exposed to the ether treatment, a systematically larger number of affected animals was seen in each generation, suggesting a cumulative effect of the treatment across generations. The response was seen equally in genetically diverse groups as well in a relatively homozygous or isogenic group, indicating that it was not restricted to a certain or highly specific genetic makeup that was more susceptible to the development of the phenotype. To show definitely that the effect was mediated by an altered cytoplasm, after the sixth generation, treated females (TF) were mated with untreated, control males (TF × UM) and treated males (TM) were bred with untreated, control females (TM × UF), with the result that only in the TF × UM group did the bithorax phenotype appear. Since cytoplasmic alterations are carried exclusively by the mother's cytoplasm and not the father's, the results of the crosses definitely ruled in the cytoplasmic nature of the effect. The coactional systems view of epigenetic development is well summarized by Mae-Wan Ho:

> There has been a recent explosion in molecular genetics research following initial breakthroughs in recombinant DNA technology. The genomic content of every organism is for the first time susceptible to being read base by base from beginning to end. Yet the first glimmerings have already yielded major surprises. Forever exorcised from our collective consciousness is any remaining illusion of development as a genetic programme involving the readout of the DNA "master" tape by the cellular "slave" machinery. On the contrary, it is the cellular machinery which imposes control over the genes. The central role of protein–protein and protein–nucleic acid interactions in the regulation of gene expression is reinforced many times over by the detailed knowledge which has recently come to light. . . . The classical view of an ultraconservative genome—the unmoved mover of development—is completely turned around. Not only is there no master tape to be read out automatically, but the "tape" itself can get variously chopped, rearranged, transposed, and amplified in different cells at different times. (Ho, 1984, p. 285)

Thus, the first component in our preliminary sketch of phenogenesis, or the developmental theory of the phenotype, is established: the genes are part of the ontogenetic system and, as such, their activity is influenced by events occurring at higher levels in the coactional system of ontogenetic development (supragenetic influences). And, in some cases, alterations of genetic activity in one generation can be carried over into succeeding generations via cytoplasmic changes that the mother passes along with her strictly genetic contribution to her offspring. It has also been experimentally demonstrated that such environmentally produced transgenerational, supragenetic, maternal contributions can influence the behavior as well as the anatomy and physiology of the grandchildren (Clark & Galef, 1988; Denenberg & Rosenberg, 1967; Huck, Labov, & Lisk, 1986; Skolnick, Ackerman, Hofer, & Weiner, 1980). This transgenerational effect is essential for the potential evolutionary significance of developmental change that does not involve mutation or genetic recombination but, rather, the expression of hitherto unexpressed genetic–cytoplasmic possibities.

Heritability

One of the major insights of the present viewpoint has to do with our understanding of the concept of heritability. While most evolutionary biologists believe there is a dichotomy in the "source" of phenotypic traits so that some traits (or the differences in some traits) are thought to be heritable and others nonheritable, the present phenogenetic viewpoint holds that genes are an inextricable component of any developmental system and thus *genes are involved in all traits*. A consequence of this view is that all traits, whether behavioral, physiological, or anatomical, are heritable and thus subject to natural selection or selective breeding—it is not the case that some traits (or differences) are heritable and others are not heritable. The straightforward prediction is that if one selectively breeds for a trait, one will promulgate that trait (or the difference in that trait), and this holds for any and all traits. Although this may seem an extreme point of view, I believe this contention can be, and has been, proven numerous times by selective breeding experiments. For example, in the 1920s a psychologist named Robert Choate Tryon continued one of the first behavioral-genetic experiments by selectively breeding rats for their ability to learn their way through a maze. Tryon selectively inbred the superior learners among each other and the poor learners among each other. Despite the fact that maze learning is a very complex behavioral trait, undoubtedly involving many complexes of genes, after a relatively small number of generations Tryon (1942) produced two strains of rats, which came to be known as "maze-bright" and "maze-dull" lines. As shown in Figure 12–4, after only seven generations of breeding the brightest rats within each of the brightest litters, and the dullest within each of the dullest litters, the performance of the two groups in maze-learning was virtually nonoverlapping. Thus, two separate strains or varieties had been established with only seven generations of inbreeding![3]

Since Tryon's time, behavioral geneticists have repeatedly demonstrated that any behavioral phenotype for which individual variation exists (e.g., emotionality, activity–inactivity) can be made the subject of selective breeding, thus supporting the phenogenetic contention that genes are involved in all phenotypic outcomes and that *all traits are heritable*. Heritable in this sense means that all behavioral traits that can be subjected to selective breeding will change in the selected direction across generations as a consequence of such selective breeding provided that the gestational and rearing conditions are kept as constant as possible across generations. A number of writers (e.g., Eisenberg, 1976; Lickliter & Berry, 1990; Mon-

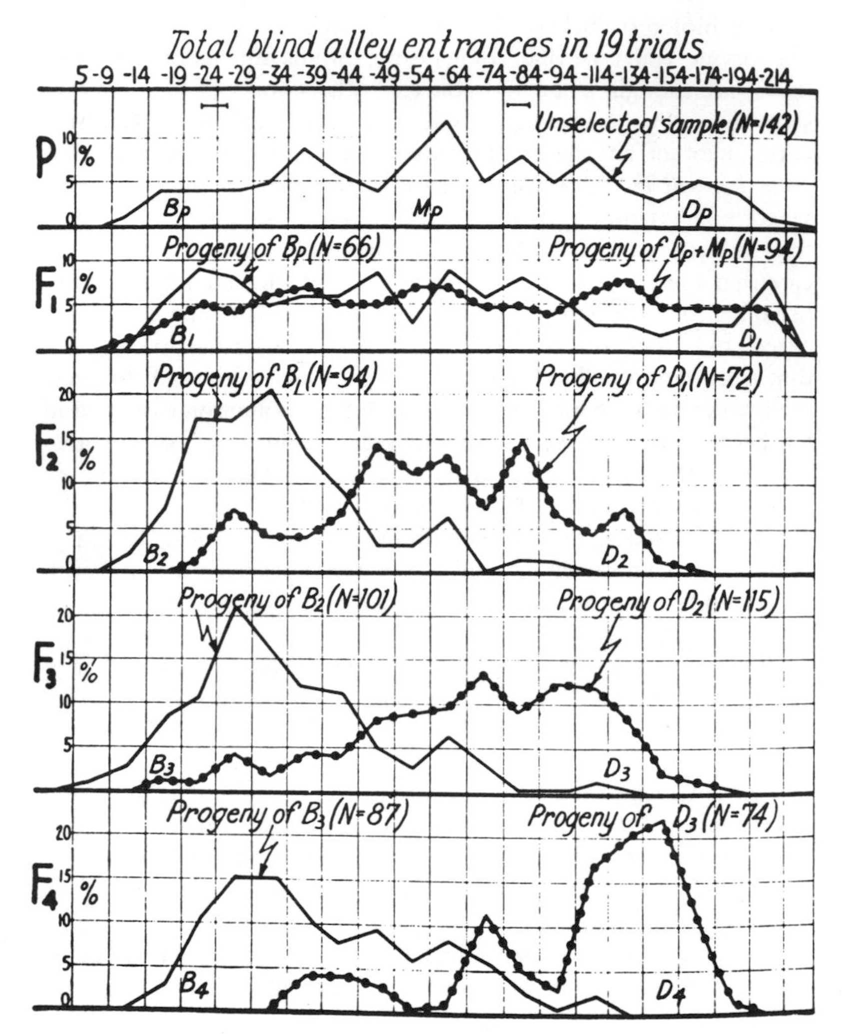

Fig. 12–4. Selective breeding for maze-learning ability in "bright" (*B*) and "dull" (*D*) rats. In the course of only seven generations of selectively breeding the maze-bright and maze-dull animals among themselves, it was possible to produce two strains of rats that showed almost no overlap in their maze-learning ability by the seventh generation. (From Tryon, 1942.)

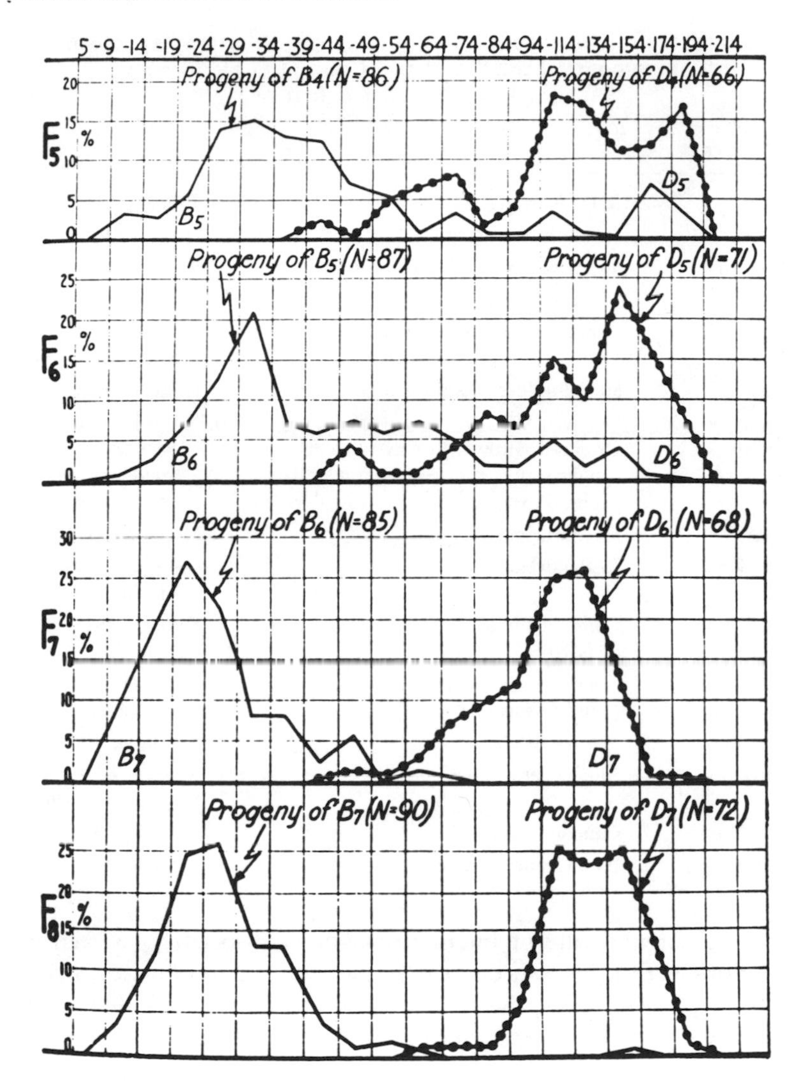

Fig. 12–4. (*continued*)

tagu, 1966; Newman, 1988; West & King, 1987) have recognized that not only genes but developmental means are necessarily transmitted between generations: "These means include genes, the cellular machinery necessary for their functioning, and the larger developmental context, which may include a maternal reproductive

TABLE 12–1. Outcomes of Behavior-Genetic Selection Experiments in Support of the Broadened Concept of Heritability Described in the Text

Species	*Behavior*	*Source*
Rats	Maze learning	Tolman, 1924
Mice	Speed in traversing runway	Dawson, 1932
Rats	Spontaneous activiy (wheel-running)	Rundquist, 1933
Rats	Maze learning	Heron, 1935
Rats	"Emotionality' in open field	Hall, 1938; Broadhurst, 1965
Mice	Sound-induced (audiogenic) seizure	Frings & Frings, 1953
Fruitflies	Orientation toward or away from light source (positive and negative phototaxis)	Hirsch & Boudreau, 1958
Rats	Saccharin preference	Nachman, 1959
Chickens	Aggressiveness	Guhl, Craig, & Mueller, 1960
Fruitflies	Components of courtship movements	Ewing, 1961
Fruitflies	Upward or downward orientation with respect to gravity (negative and positive geotaxis)	Hirsch & Erlenmeyer-Kimmling, 1961
Fruitflies	Mating speed	Manning, 1961
Mice	Alcohol preference	Rodgers & McClearn, 1962
Fruitflies	General activity	Manning, 1963
Honeybees	Nest-cleaning, stinging	Rothenbuhler, 1967

system, parental care or interactions with conspecifics, as well as relations with other aspects of the animate and inanimate world" (Oyama, 1989). Thus, the supragenetic factors are "hereditary" in the sense that the zygote, embryo, infant, etc., inherits a species-standard environment composed of parents, siblings, diet, physical factors, and so on, that contribute to the expression of the phenotype. There is no basis for the dichotomy of heritable–nonheritable because all traits arise as a consequence of ontogenetic development and genes are a part of the ontogenetic system. This is such an important aspect of the thinking underlying phenogenesis that I have compiled a list of a large variety of behavioral traits that were subjected to behavior-genetic selection experiments in the early days of the field (1924–1967). As shown in Table 12–1, behavior geneticists have shown that it is possible to select for changes in all sorts of traits: learning, spontaneous activity, seizures, alcohol preference, aggressiveness, mating speed, and so on. I am assuming the reader will agree that if behavior can be made the immediate and fast-changing subject of selective breeding, it is unnecessary to document here the same phenomenon regarding anatomy and physiology.

Since the author is not a geneticist, it is important to note that the geneticist R. Lewontin reached the same conclusion about the power of selective breeding in fruitflies, "There appears to be no character—morphogenetic, behavioral, physiological, or cytological—that cannot be selected in *Drosophila*" (Lewontin, 1974, p. 92).[4] The only reason it is possible to perpetuate virtually any trait by selective breeding in the laboratory is that great pains are taken to ensure that relevant environmental (i.e., developmental rearing) conditions remain as constant as possible over the course of the generations of selection. The fact that it is possible to select for any trait does not mean that all adaptive traits in nature have been selected for. As Professor Lewontin points out, "Selection requires two things after all: one is genetic variation, and the presence of genetic variation for a remarkable variety of traits is what the artificial selection experiments proved; second, there is the necessity of differential fitness being actually associated with the different genotypes. That is something that has not been demonstrated in nature and is a very thorny issue" (personal communication, October 23, 1989). Speaking for myself, it is clear from Table 12–1 that one can selectively breed for pathological or disadvantageous traits as well as for adaptive traits. Selective breeding is based on the outcome of differences in development. Beyond the developmental differences that produce these different outcomes, whether we are dealing with absolutely different genotypes or, more likely in my opinion, different genetic expression in much the same genotypes, remains for future research to determine.

Enormous Hidden Genetic Store for Phenotypic Variation[5]

Since genetic expression can be influenced (through the cytoplasm) from any level of the coactional developmental system shown in

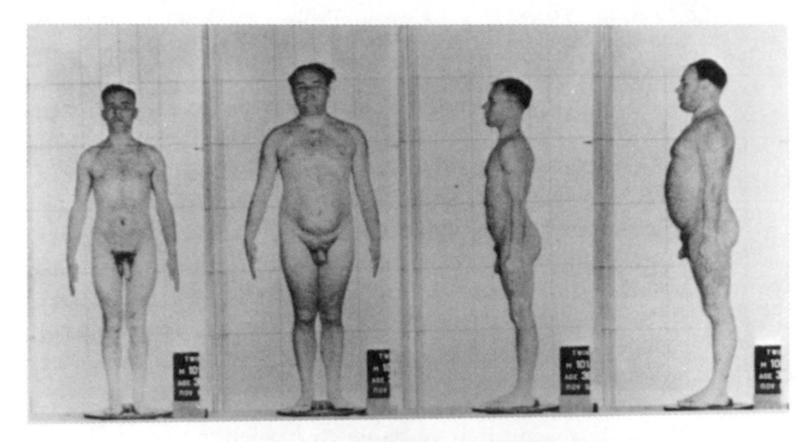

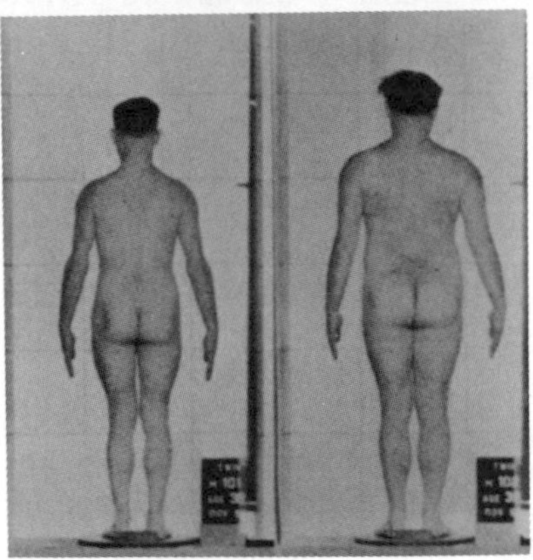

Fig. 12–5. Remarkable illustration of the enormous phenotypic variation that usually remains unexpressed in monozygotic (single-egg) identical twins reared apart. (From Tanner, 1978.)

Figures 10–5, 12–1, and 12–2, it follows from that premise that even clones or identical twins will inevitably show some variation in phenotype because organisms that have the same genotype cannot share precisely the same "experiences" or developmental pathway, even when reared under similar conditions, so somewhat different aspects of the same genetic potential will be expressed in otherwise isogenic organisms. The possible vastness of the genetic store for phenotypic variation in isogenic organisms is usually hidden because it goes unexpressed, but it is probably correct to say that

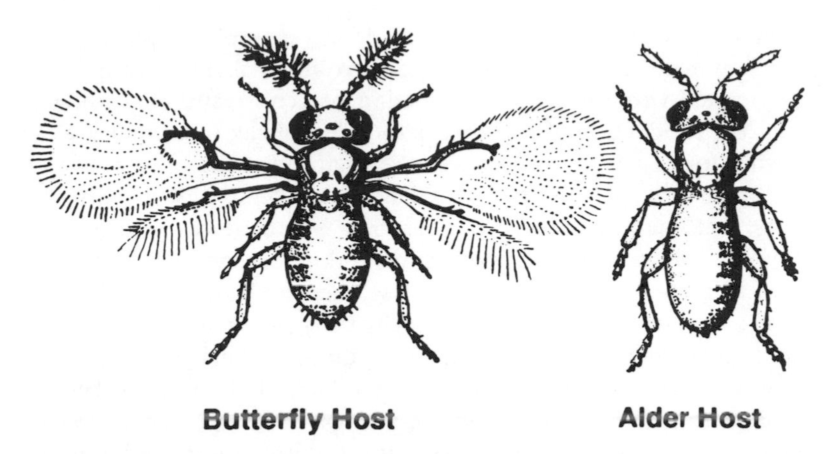

Fig. 12–6. Two very different morphological outcomes of development in the minute parasitic wasp, depending on the host in which the eggs were laid. The insects are of the same species of parasitic wasp. (Modified after Wigglesworth, 1964.)

many developmental neurobiologists were taken aback when they learned of the amount of variation in individually identified nerve cells *within clones* of such "simple" organisms as crustaceans, nematodes, fish, and grasshoppers (review and references in Goodman, 1978). To continue along the same line in an example involving human beings, certainly no one would have guessed that the two men shown in Figure 12–5 are identical twin brothers. As described by James Shields (1962, pp. 43–44, 178–180) and in Tanner (1978, p. 119), they were separated at three weeks after birth and they underwent very different rearing regimens. They exhibit well the enormous developmental potential for variation that usually remains unexpressed in closely related people.

Another striking example of the genetic and developmental potential for phenotypic variation comes from field observations of insects. The eggs of the minute parasitic wasp *Trichogramma semblidis* are sometimes laid in a butterfly host and sometimes in an alder fly host. The remarkable morphological difference in the adult phenotypes arising from these different developmental conditions can be seen in Figure 12–6. These were at one point thought to be two different species of parasitic wasps!

Finally, and perhaps most astounding with respect to hidden

genetic or developmental potential, is an example of phenotypic development in the bird (chick) embryo under highly unusual experimental conditions: the production of hen's teeth! It is known that the production of an (always toothless) beak in the chick embryo is formed as a consequence of normal tissue coactions between the avian oral epidermal tissue and avian oral mesenchyme, with the latter determining the sort of beak that will be developed. In embryological experiments, when chick oral epidermis and chick oral mesenchyme coact, as is the usual case, there appears a typical chick beak. The significance of the mesenchyme is shown when chick oral epidermis is placed in association with duck mesenchyme: the beak now shows the ridges characteristic of a duck's bill (Hayashi, 1965). The more astounding result came when Kollar and Fisher (1980) placed chick oral epidermis in contact with mouse molar mesenchyme: they obtained phenotypes of enameled dentition, otherwise known as teeth! Thus, the genetic component that is necessary for the chick oral epithelium to produce teeth has been retained from the reptilian ancestry of birds. The component that has changed during evolution is the nature of the avian oral mesenchyme, which is different from reptilian and mammalian oral mesenchyme, so, while birds do not ordinarily grow teeth during their ontogenetic development, they have the hidden genetic capability to do so.

Genes in Evolution

The preceding examples provide good evidence of the vast, almost always hidden store of genetic potentiality. That we do not yet even vaguely comprehend the nature of the contribution of the genes to development and any resulting phenotype, whether normal or abnormal, complex or simple, is further attested to by two evolutionary paradoxes: neither the size nor the complexity of the genetic complement (genome) bears any relationship to the size or complexity of organisms ranging from bacteria to humans.

As shown in Figure 12–7 and Table 12–2, there is no correlation between the complexity of organismic anatomical development and the amount or complexity of DNA in their genome. In many simpler forms (algae, mosses, salamanders, fishes, etc.), the amount

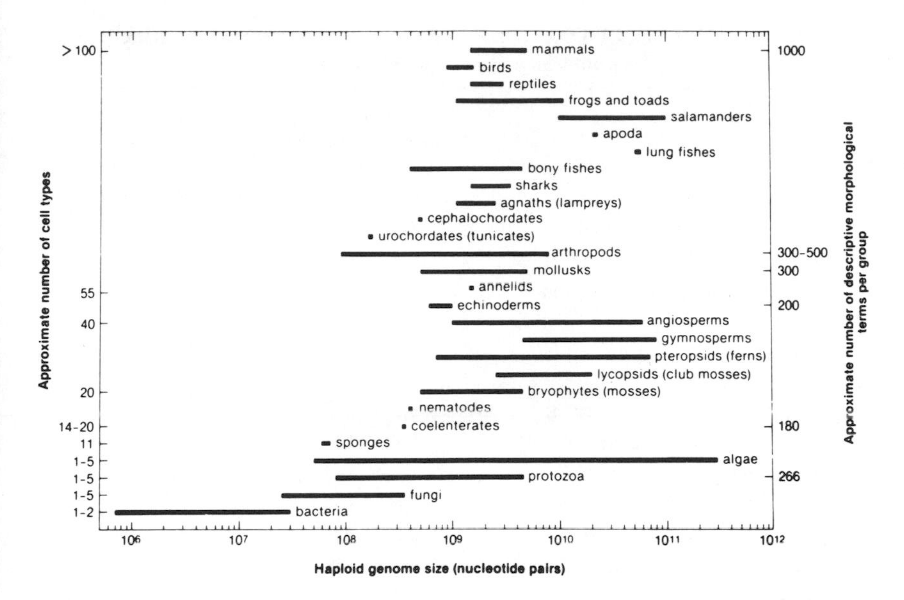

Fig. 12–7. The C-value paradox: absence of a relationship between genome size and morphological complexity. The bars show the ranges of genome sizes for various categories of organisms. The somewhat subjective ordering of categories is from morphologically most simple at the bottom to most complex at top. Two estimates of complexity are given: approximate numbers of cell types in the body of some groups are indicated on the left vertical axis and approximate numbers of morphological descriptive terms for certain groups are indicated on the right vertical axis. (C-value data from Sparrow, Price, & Underbrink, 1972. Figure from Raff & Kaufman, 1983.)

and complexity of DNA are much greater than in humans and all the other more complex species of mammals and birds. If Weismann's notion that there must be a corresponding genetic determinant for every structure of the organism (and the current versions of that idea) were to be correct, there would, of course, be a perfect or near-perfect correlation of anatomical complexity and the amount or complexity of DNA in every species. Since there is virtually no correlation between anatomical complexity and genome size or genome complexity, it is all the more clear that any and all

TABLE 12–2. DNA Diversity (Complexity) as Measured by Kinetic Genome Size (Speed of Hybridization)

Species	*DNA Content of Haploid Genome (Size)*	*Kinetic Genome Size (Complexity)*
Escherichia coli (bacterium)	4.2×10^6	4.0×10^6
Bacillus subtilis (bacterium)	3.0×10^6	3.9×10^7
Achyla bisexualis (water mold)	4.8×10^7	7.3×10^7
Dictyostelium discoideum (slime mold)	5.4×10^7	1.0×10^8
Caernorhabditis elegans (nematode worm)	8.0×10^7	9.2×10^7
Panagrellus silusiae (nematode)	8.7×10^7	2.1×10^8
Ciona intestinalis (sea squirt)	1.4×10^8	1.5×10^8
Drosophila melanogaster (fruitfly)	1.4×10^8	4.2×10^8
Bombyx mori (silk worm)	5.0×10^8	8.2×10^8
Musca domestica (house fly)	8.6×10^8	9.1×10^8
Strongylocentrotus purpuratus (sea urchin)	8.6×10^8	1.2×10^9
Physar polycephalum (mold)	9.1×10^8	7.4×10^8
Antheraea pernyi (silk moth)	9.7×10^8	1.0×10^9
Gallus domesticus (chicken)	1.2×10^9	1.1×10^9
Gecarinus lateralis (land crab)	1.3×10^9	1.8×10^9
Aplysia californica (mollusc)	1.7×10^9	2.5×10^9
Petroselinum sativum (parsley)	1.9×10^9	3.0×10^9
Cancer borealis (crab)	2.0×10^9	4.0×10^9
Mus musculus (mouse)	2.7×10^9	1.6×10^9
Rattus norvegicus (rat)	3.0×10^9	3.3×10^9
Homo sapiens (human)	3.3×10^9	1.8×10^9
Bovis domesticus (cow)	3.1×10^9	2.6×10^9
Xenopus laevis (toad)	3.1×10^9	3.0×10^9
Nicotinia tabacum (tobacco)	4.8×10^9	1.5×10^9
Bufo bufo (amphibian)	6.7×10^9	6.3×10^9
Triticum aestivum (haxaploid wheat)	1.7×10^{10}	5.2×10^9
Triturus cristatus (amphibian)	2.2×10^{10}	1.0×10^{10}
Necturus maculosis (amphibian)	5.0×10^{10}	4.2×10^{10}

From Lewin (1980) appendix 2, pp. 962–963.

The larger kinetic values signify greater DNA diversity (e.g., parsley is more genetically complex than are humans).

phenotypic features are produced by an array of coactional factors in development, of which the genes are but one part. The actual contribution of genes to individual development is still being worked out, and there seems to be good justification for the systems-like view we have adopted in this and the next chapter (Davidson, 1986).

◇ ◇ 13 ◇ ◇

From Gene to Organism: The Developing Individual as an Emergent, Interactional, Hierarchical System

The historically correct definition of epigenesis—the emergence of new structures and functions during the course of individual development—did not specify, even in a general way, how these emergent properties come into existence. Thus, there was still room for preformation-like thinking about development, which I (Gottlieb, 1970) earlier labeled the predetermined conception of epigenesis, in contrast to a probabilistic conception (see Table 13–1 for details). That epigenetic development is probabilistically determined by active interactions among its constituent parts is now so well accepted that epigenesis itself is sometimes defined as the interactionist approach to the study of individual development (e.g., Dewsbury, 1978; Johnston, 1987). That is a fitting tribute to the career-long labors of Zing-Yang Kuo (1976), T. C. Schneirla (1961), and Daniel

Fig. 13–1. ***Clockwise from upper left:*** **Zing-Yang Kuo (1898–1970), T. C. Schneirla (1902–1968), and Daniel S. Lehrman (1919–1972), principal champions of the interactionist viewpoint in psychological and behavior development.**

S. Lehrman (1970), the principal champions of the interaction idea in the field of psychology, particularly as it applies to the study of behavioral and psychological development. Thus, it seems appropriate to offer a new definition of epigenesis that includes not only the idea of the emergence of new properties but also the idea that the emergent properties arise through reciprocal interactions (coactions) among already existing constituents. Somewhat more formally expressed, the new definition of epigenesis would say that *individual development is characterized by an increase of complex-*

Table 13–1. Two Versions of Epigenetic Development

Predetermined Epigenesis
Undirectional Structure-Function Development

Genetic Activity → Structural Maturation → Function, Activity, or Experience
(DNA → RNA → Protein)

Probabilistic Epigenesis
Bidirectional Structure-Function Development

Genetic Activity ↔ Structural Maturation ↔ Function, Activity, or Experience
(DNA ↔ RNA ↔ Protein)

As applied to the nervous system, structural maturation refers to neurophysiological and neuroanatomical development, principally the structure and function of nerve cells and their synaptic interconnections. The unidirectional structure-function view assumes that genetic activity gives rise to structural maturation that then leads to function in a nonreciprocal fashion, whereas the bidirectional view holds that there are constructive reciprocal relations between genetic activity, maturation, and function. In the unidirectional view, the activity of genes and the maturational process are pictured as relatively encapsulated or insulated so that they are uninfluenced by feedback from the maturation process or function, whereas the bidirectional view assumes that genetic activity and maturation are affected by function, activity, or experience. The bidirectional or probabilistic view calls for arrows going back to genetic activity to indicate feedback serving as signals for the turning off and turning on of DNA to manufacture protein. The usual view calls for genetic activity to be regulated by the genetic system itself in a strictly feedforward manner. That the feedback (actually, feeddown) view is correct is evidenced by the experimental results of Zamenhof & van Marthens, Uphouse & Bonner, and Grouse et al. reviewed in this chapter.

Note: Throughout this work I have presented DNA → RNA → protein pathway in an oversimplified manner that, although it seems appropriate for the present purpose, does disregard the fact that a number of crucial events intervene between RNA and protein formation. In fact, according to Pritchard (1986), dozens of known factors intervene between RNA activity and protein formation! Thus, it is an oversimplification to imply that DNA and RNA alone produce specific proteins—other factors (e.g., cytoplasm) contribute to the specificity of the protein.

ity of organization—i.e., the emergence of new structural and functional properties and competencies—at all levels of analysis (molecular, subcellular, cellular, organismic) *as a consequence of horizontal and vertical coactions among its parts, including organism–environment coactions.* Horizontal coactions are those that occur at the same level (gene–gene, cell–cell, tissue–tissue, organism–organism), whereas vertical coactions occur at different levels (gene–cytoplasm, cell–tissue, behavioral activity–nervous system) and are reciprocal, meaning that they can influence each other in either direction, from lower to higher, or from higher to lower, levels of the developing system. For example, the sensory experience of a developing organism affects the differentiation of its nerve cells, such that the more experience the more differentiation and the less experience the less differentiation. (For example, enhanced activity or experience during individual development causes more elaborate branching of dendrites and more synaptic contacts among nerve cells in the brain [Greenough & Juraska, 1979.])[1] Reciprocally, the more highly differentiated nervous system permits a greater degree of behavioral competency and the less differentiated nervous system permits a lesser degree of behavioral competency. Thus, the essence of the probabilistic conception of epigenesis is the bidirectionality of structure–function relationships, as depicted in Table 13–1. It is important to note that this hierarchical, reciprocal, coactive definition of epigenesis holds for anatomy and physiology (cf. the embryologist P. D. Nieuwkoop's definition in Gerhart, 1987), as well as for behavior and psychological functioning. The traffic is bidirectional, neither exclusively bottom–up or top–down. The embryologists Ludwig von Bertalanffy (1933–1962) and Paul Weiss (1939–1969), and the geneticist Sewall Wright (1968) have long been championing such a systems view for developmental genetics and developmental biology. The systems view in developmental psychology is exemplified by approaches and theories that have been called ecological (Bronfenbrenner, 1979), transactional (Dewey & Bentley, 1949; Sameroff, 1983), contextual (Lerner & Kaufman, 1985), interactive (Johnston, 1987; Magnusson, 1988), probabilistic epigenetic (Gottlieb, 1970), individual–socioecological (Valsiner, 1987), structural–behavioral (Horowitz, 1987), and, most globally speaking, interdisciplinary developmental science (Cairns, 1979).

Developmental Causality (Coaction)

Behavioral (or organic or neural) outcomes of development are a consequence of *at least* (at minimum) *two* specific components of coaction (e.g., person–person, organism–organism, organism–environment, cell–cell, nucleus–cytoplasm, sensory stimulation–sensory system, activity–motor behavior). The cause of development—what makes development happen—is the relationship of the

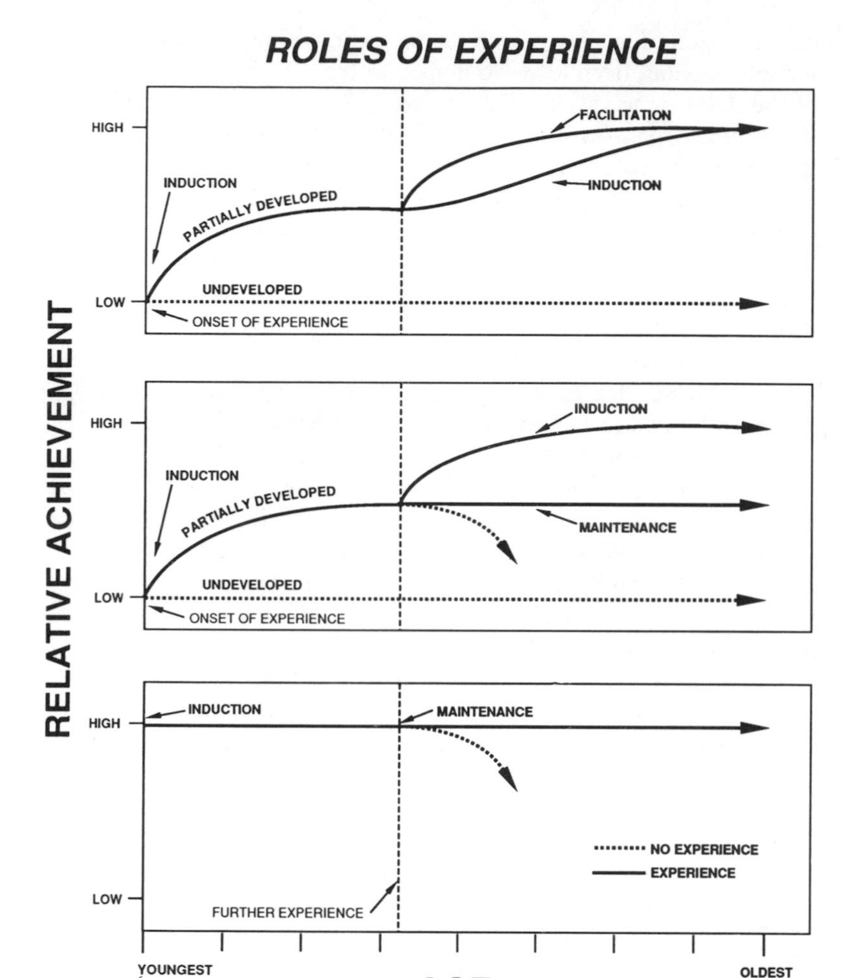

Fig. 13–2. Three roles that experience plays in the development of anatomy, physiology, and behavior.

two components, not the components themselves. Genes in themselves cannot cause development any more than stimulation in itself can cause development. When we speak of coaction as being at the heart of developmental analysis or causality what we mean is that we need to specify some relationship between at least two components of the developmental system. The concept used most frequently to designate coactions at the organismic level of functioning is *experience:* experience is thus a relational term. As documented elsewhere (Gottlieb 1976a, 1976b), experience can play at least three different roles in anatomical, physiological, and behavioral development (Figure 13–2). It can be necessary to sustain already achieved states of affairs (*maintenance function*), it can temporally regulate when a feature appears during development (*facilitative function*), and it can be necessary to bring about a state of affairs that would not appear unless the experience occurred (*inductive function*).

Since first defining and formulating the various roles of experience in 1976, it has become apparent that interactive *activity* of some kind (e.g., activity–dependent regulation of gene expression) is a usual part of development at all levels from the subcellular to the organismic, so I have here (Figure 13–2) modified the early scheme to reflect that eventuality. (See footnote 1 to this chapter for a more precise definition of intrinsic experience.) While this scheme is not exhaustive, it does call attention to the various roles that experience (interactive activity) can play in both intrinsic relations (inside the organism) and extrinsic ones (organism–organism or organism–environment relations), as Bateson (1983) and others have pointed out. In this scheme experience is defined as interactive activity, whether it is analyzed under or outside the skin of the organism, or, preferably, both under (say at the cellular level) and outside which can now be realized in research in developmental neurobiology. The techniques at hand make such thoroughgoing analyses quite feasible, especially in multidisciplinary research settings where investigators can pool their skills from the most molar levels (culture, society) to the most molecular levels of analysis. Such a multidisciplinary developmental systems approach is described and discussed in the first section of the January 1991 issue of *Developmental Psychology,* and one hopes it will become more commonplace in the years to come.

Since developing systems are by definition always changing in some way, statements of developmental causality must also include a *temporal* dimension describing when the experience occurred. For example, one of the earliest findings of experimental embryology had to do with the differences in outcome according to the time during early development when tissue was transplanted. When tissue from the head region of the embryo was transplanted to the embryo's back, if the transplantation occurred early in development the tissue differentiated according to its new surround (i.e., it differentiated into back tissue), whereas, if the transplant occurred later in development the tissue differentiated according to its previous surround so that, for example, a third eye might appear on the back of the embryo. These transplantation experiments demonstrated not only the import of time but also showed the essentially coactional nature of embryonic development.

Significance of Coaction for Individual Development

The early formulation by August Weismann (Chapter 7) of the role of the hereditary material (what came to be called genes) in individual development held that different parts of the genome or genic system caused the differentiation of the different parts of the developing organism, so there were thought to be genes for eyes, genes for legs, genes for toes, and so forth. Hans Driesch's experiment (Chapter 7), in which he separated the first two cells of a sea urchin's development and obtained a fully formed sea urchin from each of the cells, showed that each cell contained a complete complement of genes. This means that each cell is capable of developing into any part of the body, a competency that was called *equipotentiality* or *pluripotency* in the jargon of the early history of experimental embryology and *totipotency* and *multipotentiality* in today's terms (e.g., DiBerardino, 1988). Each cell does not develop into just any part of the body even though it has the capability of doing so. Each cell develops in accordance with its surround, so cells at the anterior pole of the embryo develop into parts of the head, cells at the posterior pole develop into parts of the tail end of the body, cells in the foremost lateral region of the embryo develop into forelimbs, cells in the hindmost lateral region develop hindlimbs, the dorsal area of the embryo develops into the back, and so on. Although we do not know what actually causes cells to differentiate appropriately according to their surround, we do know that it is the cell's interaction with its surround, including other cells in that same area, that causes the cell to differentiate appropriately. The actual role of genes (DNA) is not to produce an arm or a leg or fingers, but to produce protein (through the coactions inherent in the formula DNA ↔ RNA ↔ protein). The protein produced by the DNA–RNA–

cytoplasm coaction then differentiates according to coactions with other cells in its surround. Thus, differentiation occurs according to coactions *above the level of DNA–RNA coaction* (i.e., at the supragenetic level). The DNA–RNA coaction produces protein and that protein subsequently differentiates according to where it finds itself in the three-dimensional space of the embryo (the anterior-posterior, lateral, and dorsal-ventral spatial dimensions), plus the temporal dimension alluded to earlier. Thus it is the coactions during each phase of embryonic development that somehow, by means not yet understood, eventually produce a mature organism.[2]

Another demonstration that genes merely produce protein and not mature traits comes from an ingenious experiment by J. B. Gurdon (1968). If genes (DNA) in the nucleus of a cell did produce specific traits (e.g., an intestinal cell), then, should it prove possible to recover the nucleus from such a cell it would be missing that part of its genetic complement because those genes would have been used up in creating the intestinal cell. By way of proving that idea wrong, Gurdon recovered the nucleus from an intestinal cell of a tadpole and inserted that nucleus into an early embryonic (blastula) cell whose nucleus had been removed. Such nuclear transfers yield entirely normal and fertile adult male and female toads, thus proving that the differentiation of a cell does not require any loss whatsoever of genetic material and that differentiation occurs above the level of the genes after the genes, in coaction with RNA and cytoplasm, have manufactured the protein that is an essential building block of all cells in the body. When certain scientists refer to behavior or any other aspect of organismic structure or function as being "genetically determined" they are not mindful of the fact that genes synthesize protein and not fully developed features of the organism. And, as experiments on the early development of the nervous system have demonstrated, the amount of protein synthesis is regulated by neural activity itself, once again demonstrating the bidirectionality and coaction of influences during individual development (Born & Rubel, 1988).

The Hierarchical Systems View

Much has been written about the holistic or systems nature of individual development. In fact, there is no other way to envisage the manner in which development must occur if a harmoniously functioning, fully integrated organism is to be its product. In earlier chapters, figures were reproduced from the writings of the geneticist Sewall Wright (Figure 10–5 in Chapter 10) and the embryologist Paul Weiss (Figure 12–2 in Chapter 12), which portray well the major components of the developing individual as an emergent, coactional, hierarchical system. So far we have dealt with the concepts of emergence and coaction as they pertain to the development of individuals. The notion of hierarchy, as it applies to individual development, simply means that coactions occur vertically as well as horizontally in all developmental systems. All the parts of the system are capable of influencing all the other parts of the system, however indirectly that influence may manifest itself. Consonant with Sewall Wright's and Paul Weiss's depiction of the developmental system, the organismic hierarchy proceeds from the lowest level, that of the genome or DNA in the nucleus, to the nucleus in the cytoplasm in the cell, to the cell in a tissue, to the tissue in an organ, the organ in an organ system, the organ system in an organism, the organism in an environment of other organisms and physical features, the environment in an ecosystem, and so on back down through the hierarchical developmental system (review by Grene, 1987). A fascinating example of the environment-to-gene hierarchical pathway is the finding that the genes that respond to heat shock in fruitfly larvae are more apt to do so if the to-be-shocked larvae have been kept at warm temperatures rather than cold ones. The characteristic heat-shock proteins that these "heat-shock genes" produce are less evident in cold-reared larvae (Singh & Lakhotin, 1988). Another dramatic developmental effect traversing the many levels from the environment back to the cytoplasm of the cell is shown by the experiments of Victor Jollos in the 1930s and Mae-Wan Ho in the 1980s. In Ho's experiment (reviewed in Chapter 12), an extraorganismic environmental event such as a brief period of exposure to ether occurring at a particular time in embryonic development can alter the cytoplasm of the cell in such a way that a different pattern of protein is produced that eventually results in a second set of wings (an abnormal "bithorax" condition) in place of the halteres (balancing organs) on the body of an otherwise normal fruitfly. Obviously, it is very likely that "signals" have been altered at various levels of the developmental hierarchy to achieve such an outcome. (Two excellent texts that describe the many different kinds of coactions that are a necessary and normal part of embryonic development are D. J. Pritchard's [1986] *Foundations of Developmental Genetics* and N. K. Wessells's [1977] *Tissue Interactions and Development.*)

It happens that when the cytoplasm of the cell is altered, as in the experiments of Jollos and Ho, the effect is transgenerational such that the untreated daughters of the treated mothers continue for a number of generations to produce bithorax offspring and do so even when mated with males from untreated lines. Such a result has evolutionary as well as developmental significance, which, to this date, has been little exploited because the neo-Darwinian, modern synthesis does not yet have a role in evolution for anything but changes in genes and gene frequencies in evolution: epigenetic development above the level of the genes has not yet been incorporated into the modern synthesis (Futuyma, 1988; Løvtrup, 1987).[3] (In the next chapter, I will describe more fully the idea that the first step in the evolutionary pathway may sometimes not involve a genetic change or mutation.)

Another remarkable organism–environment coaction occurs routinely in coral reef fish. These fish live in spatially well-defined, social groups in which there are many females and few males. When a male dies or is otherwise removed from the group, one of the females initiates a sex reversal over a period of about two days in which she develops the coloration, behavior, and gonadal physiology and anatomy of a fully functioning male (Shapiro, 1981). Such sex reversals keep the sex ratios about the same in social groups of coral reef fish. Apparently, it is the higher ranking females that are the first to change their sex and that inhibits sex reversal in lower ranking females in the group. Sex reversal in coral reef fish provides an excellent example of the vertical dimension of developmental causality.

The completely reciprocal or bidirectional nature of the vertical or hierarchical organization of individual development is nowhere more apparent than the responsiveness of cellular or nuclear DNA itself to events originating in the external environment of the organism. The major theoretical point of this monograph is that the genes are part of the developmental system in the same sense as other components (cell, tissue, organism), so genes must be susceptible to influence from other levels during the process of individual development. DNA produces protein, cells are composed of protein, so there must be a high correlation between the *size* of cells, amount

TABLE 13–2. Sequence Complexities of Visual Cortex RNAs

Group	*No. of Separate RNA Preparations*	*No. of Hybridization Reactions at ER_0t = 211,300*	*RNA Complexity as Percentage of uDNA*
Unsutured	5	18	11.77 ± 0.59
Sutured	5	10	8.69 ± 0.19
Repeat-sutured	4	8	9.22 ± 0.86

From Grouse et al. (1979).

of protein, and quantity of DNA, and there must also be a high correlation between the *number* of cells, amount of protein, and quantity of DNA, and so there is (Cavalier-Smith, 1985; Mirsky & Ris, 1951). For our behavioral–psychological purposes, it is most interesting to focus on the developing brain, where we do indeed find the expected correlation among size and/or number of brain cells, amount of protein, and quantity of DNA (Zamenhof & van Marthens, 1978, 1979). From the present point of view, it is significant that cell size, if not cell number, in the developing rodent and chick brain is responsive to two sorts of environmental input: nutrition and sensorimotor experience. Undernutrition and "supernutrition" produce newborn rats and chicks with lower and higher quantities of cerebral DNA respectively (Zamenhof & van Marthens, 1978, 1979). Similar cerebral consequences are produced by extreme variations (social isolation, environmental enrichment) in sensorimotor experience during the postnatal period (Rosenzweig & Bennett, 1978).

Since the route from DNA to protein is through the mediation of RNA (DNA → RNA → protein), it is significant for the present theoretical viewpoint that social isolation and environmental enrichment produce alterations in the complexity (or diversity) of RNA sequences in the brains of rodents. (RNA complexity or diversity refers to the total number of nucleotides of individual RNA molecules.) A specific example of a change in RNA complexity as a consequence of normal and deprived visual experience is shown in Table 13–2. When the eyelids of kittens are sutured closed so they cannot receive visual stimulation, they show less RNA complexity in the visual cortex of the brain compared to normal (unsutured) kittens (Grouse et al., 1979). In general, environmental enrichment produces an increase in the complexity of expression of RNA sequences, whereas social isolation and environmental deprivation result in a significantly reduced degree of RNA complexity (Grouse et al., 1980; Uphouse & Bonner, 1975). These experientially produced alterations in RNA diversity are specific to the brain. When other organs are examined (e.g., liver), no such changes are found.

Nonlinear Causality

Because of the emergent nature of epigenetic development, another important feature of developmental systems is that causality is often not "linear" or straightforward. In developmental systems the coaction of *X* and *Y* often produces *W* rather than more of *X* or *Y*, or some variant of *X* or *Y*. Another, perhaps clearer, way to express this same idea is to say that developmental causality is often not obvious. In my own research, for example, I found that mallard duck embryos had to hear their own vocalizations prior to hatching if they were to show their usual highly specific behavioral response to the mallard maternal assembly call after hatching. If the mallard duck embryo was deprived of hearing its own or sib vocalizations, it lost its species-specific perceptual specificity and became as responsive to the maternal assembly calls of other species as to the mallard hen's call. To the human ear, the embryo's vocalizations sound nothing like the maternal call. It turned out, however, that there are certain rather abstract acoustic ingredients in the embryonic vocalizations that correspond to critical acoustic features that identify the mallard hen's assembly call. In the absence of experiencing those ingredients, the mallard duckling's auditory perceptual system is not completely "tuned" to those features in the mallard hen's call and they respond to the calls of other species that resemble the mallard in these acoustic dimensions. The intricacy of the developmental causal network revealed in these experiments proved to be striking. Not only must the duckling experience the vocalizations as an embryo (the experience is ineffective after hatching), the embryo must experience *embryonic* vocalizations. That is, the embryonic vocalizations change after hatching and no longer contain the proper ingredients to tune the embryo to the maternal cell (Gottlieb, 1985).

Prenatal nonlinear causality is also nonobvious because the information, outside of experimental laboratory contexts, is usually not available to us. For example, the rate of adult sexual development is retarded in female gerbils that were adjacent to a male fetus during gestation (Clark & Galef, 1988). To further compound the nonobvious, the daughters of late-maturing females are themselves retarded in that respect—a transgenerational effect!

In a very different example of nonobvious and nonlinear developmental causality, Cierpial and McCarty (1987) found that the so-called spontaneously hypertensive (SHR) rat strain employed as an animal model of human hypertension is made hypertensive by coacting with their mothers after birth. When SHR rat pups are suckled and reared by normal rat mothers after birth they do not develop hypertension. It appears that there is a "hyperactive" component in SHR mothers' maternal behavior that causes SHR pups to develop hypertension (Myers, Brunelli, Shair, Squire, & Hofer, 1989; Myers, Brunelli, Squire, Shindeldecker, & Hofer, 1989). The highly specific coactional nature of the development of hypertension in SHR rats is shown by the fact that normotensive rats do not develop hypertension when they are suckled and reared by SHR mothers. Thus, although SHR rat pups differ in some way from normal rat pups, the development of hypertension in them nonetheless requires an interaction with their mother; it is not an inevitable outcome of the fact that they are genetically, physiologically, and/or anatomically different from normal rat pups. This is a good example of the *relational* aspect of the definition of experience and developmental causality offered earlier in this chapter. The cause of the hypertension in the SHR rat strain is not in the SHR rat pups or in the SHR mothers but in the nursing relationship between the SHR rat pups and their mother.

Another example of a nonlinear and nonobvious developmental experience undergirding species-typical behavioral development is Wallman's (1979) demonstration that if chicks are not permitted to see their toes during the first two days after hatching, they do not eat or pick up mealworms as chicks normally do. Instead, the chicks stare at the mealworms. Wallman suggests that many features of the usual rearing environment of infants may offer experiences that are necessary for the expression of species-typical behavior.

The Unresolved Problem of Differentiation

The nonlinear, emergent, coactional nature of individual development is well exemplified by the phenomenon of *differentiation,* whereby a new kind of organization comes into being by the coaction of preexisting parts. If genes directly caused parts of the embryo rather than producing protein, there would be less of a problem in understanding differentiation. Since the route from gene to mature structure or organism is not straightforward, differentiation poses a significant intellectual puzzle, as recognized as early as 1962 by Ephrussi (1979), among others. The problem of differentiation also involves our limited understanding of the role of genes in development; what else, if anything, might genes do than produce protein?

It has been recognized since the time of Driesch's earth-shaking experiments demonstrating the genetic equipotentiality of all cells of the organism that the chief problem of understanding development was that of understanding why originally equipotential cells actually do become different in the course of development, i.e., how is it they differentiate into cells that form the tissues of very different organ systems. The problem of understanding development thus became the problem of understanding cellular differentiation. We still do not understand differentiation today, and it is quite telling of the immense difficulty of the problem that today's theory of differentiation is very much like the necessarily vaguer theories put forth by E. B. Wilson in 1896 and T. H. Morgan in 1934 (reviewed in Davidson, 1986), that ultimate or eventual cellular differentiation is influenced by an earlier coaction between the genetic material in the nucleus of the cell with particular regions of the cytoplasm of the cell. Some of the vagueness has been removed in recent years by the actual determination of regional differences in the cytoplasm (extensively reviewed by Davidson, 1986). Thus, the undifferentiated protein resulting from locale or regional differences of nucleo–cytoplasmic coaction is biochemically distinct, which, in some as yet unknown way, influences or biases its future course of development. For example, protein with the same or similar biochemical makeups may stay together during cellular migration during early development and thus eventually come to form a certain part of the organism by the three-dimensional spatial field considerations of the embryo mentioned earlier in this chapter. Although the actual means or mechanisms by which some cells become one part of the organism and others become another part are still unresolved, we do have a name for the essential coactions that cause cells to differentiate: they are called embryonic *inductions*. The nonlinear hallmark of developmental causality is well exemplified by embryonic induction, in which one kind of cell (*A*) coacting with a second kind of cell (*B*) produces a third kind of cell (*C*). For example, if left in place, cells in the upper one-third of an early frog embryo differentiate into nerve cells: if removed from that region, those same cells can become skin cells. Equipotentiality and the critical role of spatial position in determining differentiation in the embryo is well captured in a quotation from the autobiography of Hans Spemann, the principal discoverer of the phenomenon of embryonic induction: "We are standing and walking with parts of our body which could have been used for thinking if they had been developed in another position in the embryo" (translated by B. K. Hall, 1988, p. 174). It might have been even more striking—and equally correct—if Spemann had elected to say, "We are sitting with parts of our body which could have been used for thinking . . . "!

Even if we do not yet have a complete understanding of differentiation, the facts at our disposal show us that epigenetic development is correctly characterized as an emergent, coactional, hierarchical system that results in increasingly complex organization. It remains now to use that conceptual framework to fashion a developmentally based view of evolutionary change.

◇ ◇ 14 ◇ ◇

Induction of Behavioral Change in Individual Development as Prelude to Evolution: The Supragenetic Developmental Basis of Evolutionary Change

The extreme malleability or plasticity of cells early in their development is mirrored to a certain, if lesser, degree in the psychological, behavioral, and neural functioning of the developing organism. The early developmental adaptability of organisms has significance for our understanding of evolution that is different from the "genes-for-traits" view that is a fundamental assumption of the population-genetic underpinning of the modern synthesis (Chapters 10 and 11). It is the purpose of the present chapter to make a case for the extragenetic or, better, the supragenetic developmental basis of evolutionary change through the genesis of novel behavioral phenotypes. To make things as clear as possible, I will contrast this developmental approach to evolution with the population-genetic model of the modern synthesis. I should say at the outset that the present theory can be integrated with the population-genetic model, with the exception of the radically different role ascribed to genes in the two viewpoints.

The Relationship of Development to Evolution

As noted in an earlier chapter, the concept that changes in individual development are the basis for evolution was raised originally by St. George Mivart (1871) in his book *On The Genesis of Species.* William Bateson (1894) favored the idea that phenotypic variation was developmentally inspired but little was done to further work out the details until Walter Garstang (1922) and Gavin de Beer (1930, 1958) delivered their respective coups de grâce to Ernst Haeckel's recapitulation doctrine, and, from another side entirely, Richard Goldschmidt (1933, 1952) hypothesized that changes in early embryonic development would be necessary for evolution to occur. While Garstang and de Beer were interested in showing the importance of various kinds of ontogenetic changes to evolution generally, Goldschmidt, having become convinced of the impossibility of neo-Darwinian microevolution producing a new species, had come to view developmental macromutation as essential to the production of the large differences necessary for speciation.

The foregoing scientists were the principal ones to establish the developmental basis of evolutionary change. The view that I wish to propose in this chapter builds on their pioneering insight, but it is different in a very important way. To be specific, Garstang, de Beer, and Goldschmidt, in agreement with the proponents of the modern synthesis, believed that a genetic change or mutation is necessary to bring about the developmental changes that lead to evolution. The point I wish to advocate is that there is so much untapped potential in the existing developmental system (including the genes) that evolution can occur without changing the genetic constitution of a population. Such changes may eventually lead to a change in genes (or gene frequencies) but evolution will have already occurred at the phenotypic level before the genetic change, if it does eventually occur, does occur. According to the present viewpoint, genetic change is a secondary or tertiary consequence of enduring behavioral changes brought about initially by supragenetic alterations of normal or species-typical development.

The Induction of Behavioral Neophenotypes

In a book closing out his underappreciated but otherwise illustrious research career as a broadly based developmental scientist, Zing-Yang Kuo (1976) coined the term *behavioral neophenotype* to refer to momentous behavioral changes or deviations from normality that could be brought into existence by altering the usual conditions of an animals' early development or experience. Kuo's purpose for advocating the creation of behavioral novelties was to show that species-typical behavior was rather more highly modifiable than anyone believed and not rigidly or narrowly fixed by genetic, innate, or instinctive constraints. To make his point, Kuo did such things as "create" a male dog that had no reproductive interests in female dogs in heat and, further, actively prevented other males from engaging in sexual behavior with such females. Kuo chose to tamper with reproductive behavior to make his point all the more telling: the usual or normal behavioral predilection for male dogs to copulate with females in heat (and thereby perpetuate the species) is a result of their having been exposed to usual or normal developmental conditions, not to instincts dictated by their genetic endowment. Kuo's argument that the establishment of behavioral neophenotypes by altering developmental circumstances should be one of the major aims of experimental animal psychology has not caught on because it seemed to many to be at best nonbiological and at worst scatological. I hope that by now supplying a broader rationale the significance of behavioral neophenotypes will be better appreciated.

If my logical arguments and presentation of the evidence to this point have been persuasive, the reader should be convinced that internal and external coactions during individual development create the resulting phenotype. This will, of course, be just as true for behavior as for anatomy or physiology. According to the viewpoint being developed in this chapter, the ease or difficulty of creating behavioral neophenotypes, and the directions of most ready behavioral change, would allow us to assess the immediately present evolutionary potential of a species. Naturally, we will expect to find that some species possess much greater immediate behavioral malleability or plasticity than other species, and that in itself will be informative about the pace and range of immediate evolutionary potential in those species.

But I am getting a little bit ahead of my story. I am certainly not being original in suggesting that behavioral innovations lead the way to evolutionary change. A number of biologists of various persuasions have resuscitated Lamarck's notion of the centrality of behavioral change to evolution (e.g., Bonner, 1983; Hardy, 1965; Larson, Prager, & Wilson, 1984; Leonovicová & Novák, 1987; Mayr, 1982; Piaget, 1978; Plotkin, 1988; Reid, 1985; Sewertzoff, 1929; Wyles, Kunkel, & Wilson, 1983). What is new and not yet widely appreciated is the supragenetic means (neophenogenetic pathway described later) by which normal or usual development can be altered so as to produce a behavioral neophenotype that is likely to lead to evolutionary change. (In an excellent chapter on this topic, P. Bateson [1988] has come to much the same conclusion.) In essence, as is widely recognized, what needs to happen is the production of animals that live differently from their forebears. Living differently, especially living in a different place, will subject the animals to new stresses, strains, and adaptations that will eventually alter their anatomy and physiology (without necessarily altering the genetic constitution of the changing population). The new situation will call forth previously untapped resources for anatomical and physiological change that are part of each species' already existing developmental adaptability. At some time further down the road it is possible the genetic makeup of the evolving population may change, but by the time that happens (if it does) the new behavioral, anatomical, and physiological changes will already be in place. The neophenogenetic pathway for evolutionary change is thus seen as (1) an alteration of development leading to a significant change in behavior, followed by (2) a change in morphology, and, eventually, possibly (3) a change in genetic composition of the population. Consistent with their view of the strictly genetic determination of the phenotype, adherents of the modern synthesis would consider that evolution occurred only if and when step (3) was achieved. From the present point of view, enduring transgenerational changes in behavior and morphology (i.e., phenotypic evolution) have occurred by step (2), without the necessity of adding to,

Table 14–1. Three Possible Stages in Evolutionary Pathway Initiated by Behavioral Neophenotype

	Evolutionary Pathway	
I: Change in Behavior	*II: Change in Morphology*	*III: Change in Genes*
First stage in evolutionary pathway: change in ontogenetic development results in novel behavioral shift (behavioral neophenotype), which encourages new environmental relationships.	Second stage of evolutionary change. New environmental relationships bring out latent possibilities for morphological-physiological change. Somatic mutation or change in genetic regulation may also occur, but a change in structural genes need not occur at this stage.	Third stage of evolutionary change resulting from long-term geographic or behavioral isolation (separate breeding populations). It is important to observe that evolution has already occurred phenotypically before stage III is reached. Modern neo-Darwinism, however, does not consider evolution to have occurred unless there is a change in genes or gene frequencies.

subtracting from, or otherwise changing the original genetic composition of the population. The present view holds that genes are part of a very flexible and highly adaptable developmental system, but that genes do not determine the features of the mature organism. Consequently, from this point of view, evolution involves changes in the developmental system (of which the genes are an essential part), but not necessarily changes in the genes themselves. For instance, it is entirely consistent with the present proposal that alterations in development may cause genes to become active in the developmental process that were heretofore quiescent. It is well accepted among developmental geneticists that only a very small portion of the genome is expressed during individual development, so there is always present a large untapped genetic resource that can be brought to surface under abnormal (non-species-typical) developmental circumstances, whether internal or external to the organism. The behavioral neophenogenetic pathway of evolutionary change is depicted in Table 14–1.

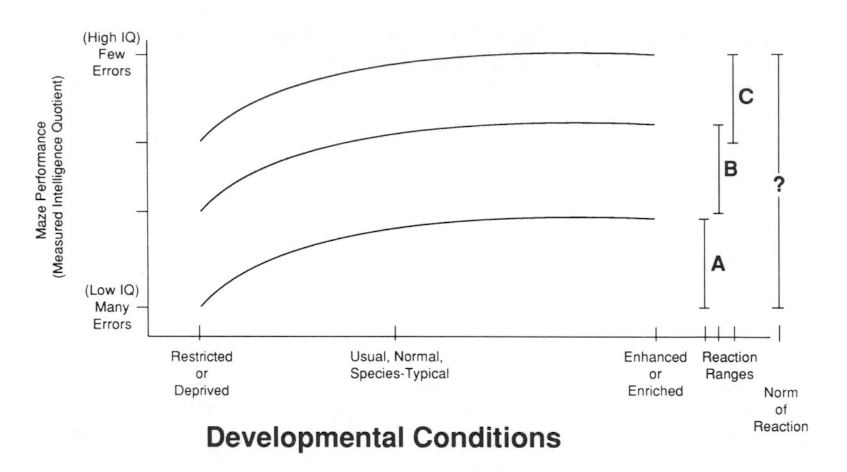

Fig. 14–1. Contrasting predictions of reaction range and norm of reaction for maze performance or IQ of three genotypes or individuals, as a consequence of being reared under three developmental conditions. The reaction-range concept predicts a narrow range of variation for each individual over the three rearing conditions, whereas the norm of reaction predicts an indefinitely broader range of phenotypic development.

Norm of Reaction versus Reaction Range

The genetic concept that I am appealing to here is really not as radical as it may sound. I am appealing to the well-known but not often appropriately used notion of a "norm of reaction," which stands in marked contrast to the better-known and more often used notion of a "reaction range." To quote from a recent article by two behavioral geneticists:

> The norm-of-reaction refers to all phenotypic outcomes of a single genotype exposed to all possible environments. It recognizes both the theoretically possible and experimentally measured outcomes and presupposes no practical limits on phenotypical variability. The reaction-range concept presumes that the genotype imposes a priori limits (a range) on the expression of a phenotype. (Platt & Sanislow, 1988)

Hypothetical behavioral–psychological examples of the reaction-range and norm-of-reaction concepts are presented in Fig-

TABLE 14–2. Mean Errors in Hebb–Williams Maze of Maze-Bright and Maze-Dull Rat Strains Reared under Enriched and Restricted Circumstances

Strain	*Enriched Rearing*	*Restricted Rearing*
Bright	111.2	169.7
Dull	119.7	169.5

From Cooper & Zubek (1958).

No statistically significant differences between the groups in either rearing condition.

ure 14–1. As defined by Gottesman (1963), an individual's heredity or genotype fixes a relatively narrow reaction range within which the behavioral phenotype will fall over a wide variety of rearing or developmental conditions. As defined by Platt and Sanislow, following Schmalhausen (1949) and Dobzhansky (1955), the norm of reaction describes a broader and less definite range of outcomes for the same genotypes when their performance is examined over a variety of developmental conditions. The latter concept is the one utilized in the present account.

An excellent illustration of the norm-of-reaction concept is provided by the work of Cooper and Zubek (1958). They reared maze-bright and maze-dull rats in either an enriched environment or a restricted environment and then tested them in the very difficult Hebb–Williams maze (Figure 14–2a,b). Since Cooper and Zubek had the narrow reaction-range concept in mind when they performed the experiment, they thought that the learning of both the bright and dull rats would improve relative to each other under the enriched rearing circumstance and would be poorer relative to each other when reared under the restricted condition. Instead, as shown in Table 14–2, what they found was equality of performance under both rearing conditions! The "dull" rats made as few errors as the "bright" rats after enriched rearing and the "bright" rats made as many errors as the "dull" rats after restricted rearing.

When the so-called bright and dull rats are tested in the Hebb–Williams maze after being reared in their usual or normal way (not enriched or restricted), there are of course significant differences between the strains (see Table 14–3) because this developmental

TABLE 14–3. Mean Errors in Hebb–Williams Maze-Bright and Maze-Dull Rat Strains Reared under Normal Conditions

Strain	*Normal (Usual Rearing Condition)*
Bright	117.0
Dull	164.0

From Hughes & Zubek (1956).
Significantly different means between the groups.

situation repeats the rearing conditions under which the original selective breeding for superior and inferior performance was carried out. If the genes coded for learning ability or established a delimited reaction range, then when these rat strains were reared under enriched or restricted conditions, the relative difference between them should be preserved. Instead, what the experiments show is that the genes are part of the developmental system, and it was the highly specific consequences of rearing under a certain developmental condition that were realized by selection. As called for by the norm-of-reaction concept, selective breeding under one developmental regimen does not predict what the animals will do when reared under different developmental conditions. The results of selection depend on the entire developmental manifold, not merely on the genes that are involved.

Determinants of Behavioral Plasticity

The present theory lays great store in the malleability or adaptability of organisms, especially the higher vertebrates (birds and mammals—more on this latter point later). The creation of behavioral neophenotypes is necessarily dependent upon the existence of some degree of behavioral plasticity or adaptability. Thus, the determinants of behavioral plasticity are an important consideration. One key limiting component of plasticity is the nervous system, particularly the brain, and the other is the developing organism's early experiences. These two components are in lockstep: larger-brained species can make more of their early experiences and early experiences affect the maturation and size of the brain. Thus, the most conspicuous developmental route to increasing behavioral plasticity and creating behavioral neophenotypes is through early experiential alterations (including nutrition) that have positive effects on enhancing the maturation of the brain.

Beginning in the 1950s, developmental psychobiologists began in earnest to study the influence of early rearing experiences on enhancing the nervous system and later exploratory behavior and problem-solving ability, the latter two interrelated forms of behavior and psychological functioning being of most relevance in engendering the sort of evolutionary progression described previously and depicted in Table 14–1. For the present purposes, we are going to be most interested in the developmental conditions that produce the sorts of behavioral plasticity that would enhance the likelihood of an individual (1) being able to survive by behavioral means in a drastically changed environment or (2) one whose behavior would be likely to bring it into a new environment, thus precipitating the anatomical and physiological changes in stage II in Table 14–1. Led by the pioneering experiments of Seymour Levine (1956) and Victor Denenberg (1969), a large number of studies showed that the unusual, perhaps stressful, experience of subjecting young rodents to handling by human beings during early development resulted in producing relatively stress-resistant animals, ones who would be capable of exploration (instead of freezing) and adaptive learning when faced with a completely strange and unfamiliar environment in adulthood.[1] The research of Levine (1962) and his collaborators showed that the axis between the adrenal and pituitary glands was enhanced by the handling experience and this anatomical–physiological change was correlated with the handled animal being able to tolerate greater stress in adulthood. As shown by Denenberg (1964) and his colleagues, the handling experience had to occur early in development if it was to be effective. Animals subjected to the same experience at older ages did not benefit from the experience, as indicated by later tests of resistance to stress and of exploratory behavior.

In a series of experiments with a rather different purpose, a doctoral student named Bernard Hymovitch (1952), under the guidance of his mentor, Donald Hebb (1947, 1949), showed in a definitive manner how variations in early experience are crucial to later *problem-solving* in adulthood. He reared young rats under

Fig. 14–2a. Hebb–Williams Closed-Field Maze. After the animal learns the way from the start to the goal box in this versatile maze, the animal is given a series of changing problems to solve in the same arena. See Figure 14–2b.

four conditions and then later tested them in the Hebb–Williams (1946) maze (Figure 14–2). The animals were housed individually in (1) a stovepipe cage (which permitted little motor or visual experience), (2) an enclosed running or activity wheel (which permitted a lot of motor activity but little variation in visual experience), (3) a mesh cage that restricted motor activity but allowed considerable variation in visual experience as it was moved daily to different locations in the laboratory. (4) The fourth group of animals contained twenty animals that were reared socially in a so-called free environment box that was very large (6′ × 4′) compared to the other conditions, and was fitted with a number of blind alleys, inclined runways, small enclosed areas, apertures, etc., that offered the rats a wide variety of opportunities for motor and visual exploration and learning in a complex physical environment. The animals lived in these four environments from about 27 days of age to 100 days of age, at which time testing in the Hebb–William maze was completed. The results of testing are shown in Table 14–4.

As shown in Table 14–4, rearing in the stovepipe and the enclosed running wheel led to the same level of poor performance,

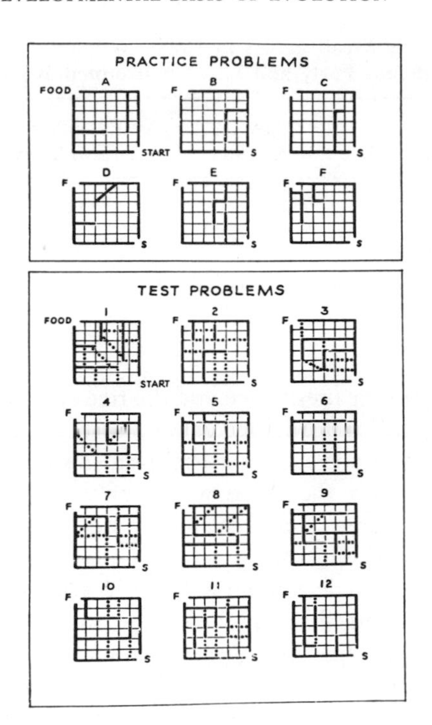

Fig. 14–2b. Problems in Hebb–Williams Maze. These are illustrations of the numerous practice and test problems that the animal encounters in this maze. The essential feature is that the path between the start and finish boxes is altered from problem to problem by moving the walls as indicated. This maze is considerably more difficult than a Y- or T-maze, so it taxes the animal's learning ability to a much greater degree than the usual maze. (From Rabinovitch & Rosvold, 1951.)

TABLE 14–4. Mean Errors in Hebb–Williams Maze of Rats Reared under Four Different Conditions.

Stovepipe	*Running Wheel*	*Mesh Cage*	*Free Environment*
223	235	140	137

From Hymovitch (1952).

The Stovepipe and Running Wheel groups made significantly more errors than the Mesh Cage and Free Environment groups.

TABLE 14–5. Mean Errors in Hebb–Williams Maze of Rats with Different Early and Late Environmental Experiences

Free Environment/ Stovepipe/	*Stovepipe/ Free Environment*	*Free Environment/ Free Environment*	*Normal Cage/ Normal Cage*
161	248	152	221

From Hymovitch (1952).

The Stovepipe/Free Environment and Normal Cage groups made significantly more errors than the other two groups.

whereas rearing in the mesh cage and the free environment led to the same level of good performance over twenty-one days of testing in the Hebb–Williams maze. All the groups also showed the same level of improvement over the three weeks of testing, so the animals reared in the mesh cages and free environment began functioning at a superior level early in testing.

Next, in order to determine whether it was the early experience in each environment that made for the differences between the groups, Hymovitch repeated the experiment with four groups of animals that differed in *when* they had the free-environment or stovepipe experience: One group had the free-environment experience from thirty to seventy-five days of age and then were placed in the stovepipe for forty-five days; a second group had the stovepipe experience from thirty to seventy-five days and then had the free-environment experience for forty-five days; a third group remained in the free environment throughout the experiment; and a fourth group remained in their normal laboratory cages throughout the experiment (these would be the most thoroughly or consistently deprived from the standpoint of motor and visual experience).

As can be seen in Table 14–5, the animals that experienced the free environment early and the stovepipe later in life performed just as well as the animals that remained in the free environment throughout the experiment. The crucial finding is that the animals who experienced the stovepipe environment early and the free environment later in life performed as poorly as the animals that remained in their normal cages throughout the experiment (the most deprived group). It is important to note that these differences in problem-solving ability were not in evidence when Hymovitch challenged the rats with a simpler, alley maze, more like the ones that were in wide use in most animal learning laboratories at the time. It is only when they were challenged by the much more difficult Hebb–Williams series of problems that the differences in problem-solving ability were in evidence.

Forgays and Forgays (1952), other students of Donald Hebb's, undertook to replicate Hymovitch's important findings and also to determine (1) whether the "playthings" in the free environment were crucial and (2) why the mesh-cage-reared animals did so well without direct experience of interacting with the multifarious objects in the free environment. They found indeed that presence of the "playthings" (inclined planes, blind alleys, etc.) were essential to the superior performance of the free-environment animals and that the mesh-cage-reared animals only do as well when their cages are moved about frequently so that they visually encounter a considerable degree of varied environmental input, including the opportunity to watch the animals in the free-environment with the playthings.

It was not long before these early experience studies were extended to other animals, including nonhuman primates, where social isolation and otherwise highly restricted, deprived rearing conditions were employed. Indeed, even in primates with relatively large brains, the normal or usual variety of experiences early in life was critical for the appearance of normal exploratory and learning abilities later in life. Deprived infants showed severe deficiencies in their later behavior (Harlow, Dodsworth, & Harlow, 1965). Just having a large brain is insufficient for the development and manifestation of the superior problem-solving skills characteristic of primates (Mason, 1968; Sackett, 1968).[2]

Thus, behavioral plasticity that is essential to behavioral neophenogenesis is dependent upon variations in early experience as well as possessing a large brain. The fully bidirectional developmental system of genetic, neural, behavioral, and environmental influences is shown in Figure 14–3, which is a much simplified and abbreviated version of Sewall Wright's developmental system shown in Figure 10–5. The conditions that favor the appearance of a behavioral neophenotype are severe or species-atypical alterations in environmental contingencies early in life. These changed contingencies can arise in two ways in animals living in nature: (1) some sort of physical or geographical change happens *to* (is forced on) the animal (a disruption of habitat, climatic change, and so on) and,

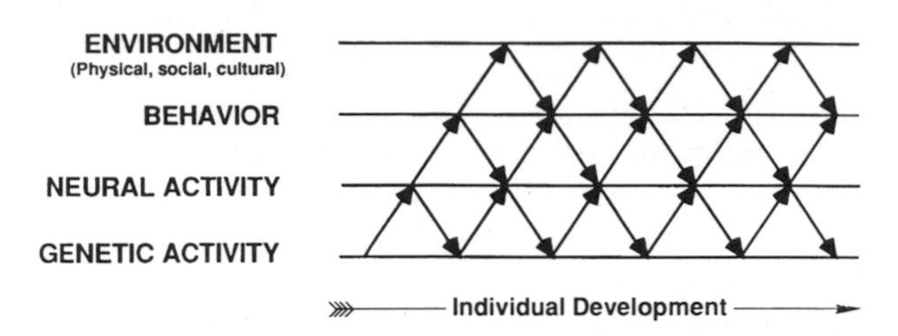

Fig. 14–3. Depiction of the completely bidirectional nature of genetic, neural, behavioral, and environmental influence over the course of individual development.

probably more frequently, (2) the migration of the animal into a somewhat different habitat based on normal exploratory behavior. The large-brained animals that are more likely to withstand (1) and commit (2) are ones that have had not only traditional but non-traditional variations in their early experience. To put it the other way around, exposure to conservative or narrow social and physical environmental contingencies early in development will make animals less likely to withstand (1) and unlikely to perpetrate (2). These predictions on evolutionary readiness, as it were, follow from the results of the early experience studies reviewed previously. There is a developmental dynamic that causes animals to prefer the familiar and thus to strive to reinstate earlier life situations or repeat versions of their early life experiences in adulthood. Consequently, it could be that animals that have had considerable variation in their early social and physical experiences will tend to seek out such variation in adulthood—just what is needed to heighten exploratory behavior and encourage novelty-seeking! While actual developmental experiments have not yet been done to show that animals (including humans) that have had considerable variation in their early experience will tend to seek out novel experiences as adults, there are two studies of adult mammals and birds that show that novelty is a psychological dimension of experience that can be abstracted such that animals so trained will consistently prefer to interact with novel rather than familiar objects or situations when given a choice (Honey, 1990; Macphail & Reilly, 1989). From the present theoretical standpoint, it would be most valuable to validate

TABLE 14–6. Modes of Behavioral Neophenogenesis

Unusual (e.g., "handling") and enriched early experiences lead to

1. Increased resistance to stress
2. Increased brain size
3. Enhanced exploratory behavior
4. Enhanced problem-solving (learning ability) in adulthood

The above would aid adaptation should (1) the organism's usual environment change drastically and (2) would also support the seeking out of new habitats in the absence of environmental change. (1) and (2) are often invoked as the initial stages of evolution.

the developmental induction of novelty-seeking behavior in later life through the experience of considerable variation early in life.

Another way, albeit indirect, to test the behavioral neophenogenetic hypothesis about evolutionary readiness is to examine the exploratory behavior and rate of evolutionary change in large-brained versus smaller-brained species. The cerebral component of behavioral neophenogenesis predicts a higher degree of exploratory behavior in large-brained species versus small-brained species and a consequent faster rate of evolutionary change in larger- versus smaller-brained species. The modes of behavioral neophenogenesis reviewed thus far are summarized in Table 14–6.

Exploratory Behavior and Rate of Evolutionary Change in Large- versus Small-Brained Species

The evolutionary lineages of the five classes of vertebrates (fish, amphibian, reptile, bird, mammal) are shown in Figure 14–4. At the base of the branch are the earliest vertebrates (primitive fishes, primitive amphibians) from which present-day fishes and amphibians have evolved. Further up the branch, the ancient, extinct stem reptiles gave rise to two different lines of evolution, one culminating in present reptiles and birds (through the extinct dinosaurs, *Archaeopteryx,* etc.), and the other culminating in three kinds of mammals (placentals, pouch-young marsupials, and the egg-laying monotremes, two species of which are still with us [the spiny anteater and the duckbilled platypus].)

The vertebrate lineage has been further refined in a psychologi-

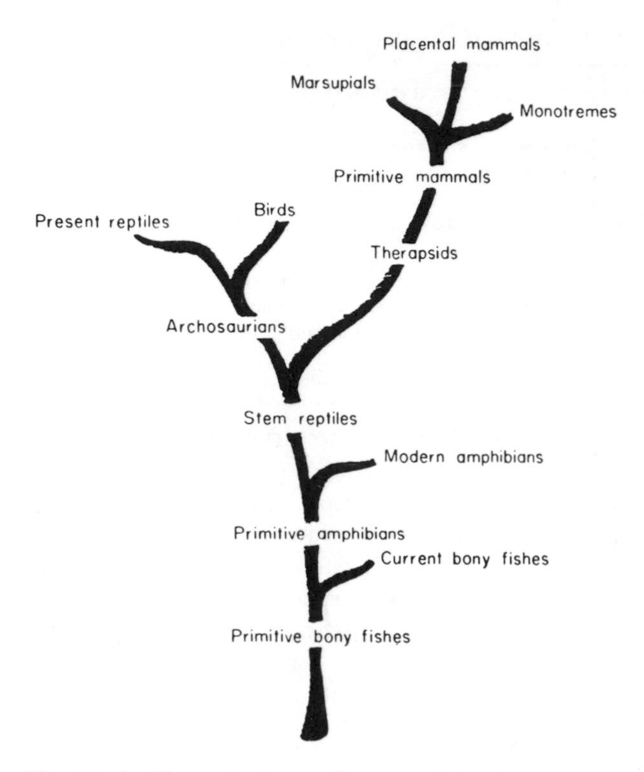

Fig. 14–4. The evolutionary lineages of the five classes of vertebrate animals: fish, amphibians, reptiles, birds, and mammals.

cally and behaviorally interesting way by Julian Huxley (1957). He discriminated between *clades* and *grades* of evolution. A clade is a delimitable, monophyletic (i.e., genetically closely related) unit produced by cladogenesis (e.g., birds or mammals, or within these large clades, songbirds or primates, respectively). A grade, on the other hand, is a particular level in an ascending series of improvements on any given structural or functional unit of analysis in animal groups that need not be closely related from a genetic point of view. Such structural and functional units, among others, would be brain:body ratio, adaptive behavior, exploratory behavior, level of problem-solving or learning ability. To exemplify the contrast between clade and grade: whereas birds and mammals are not within the same

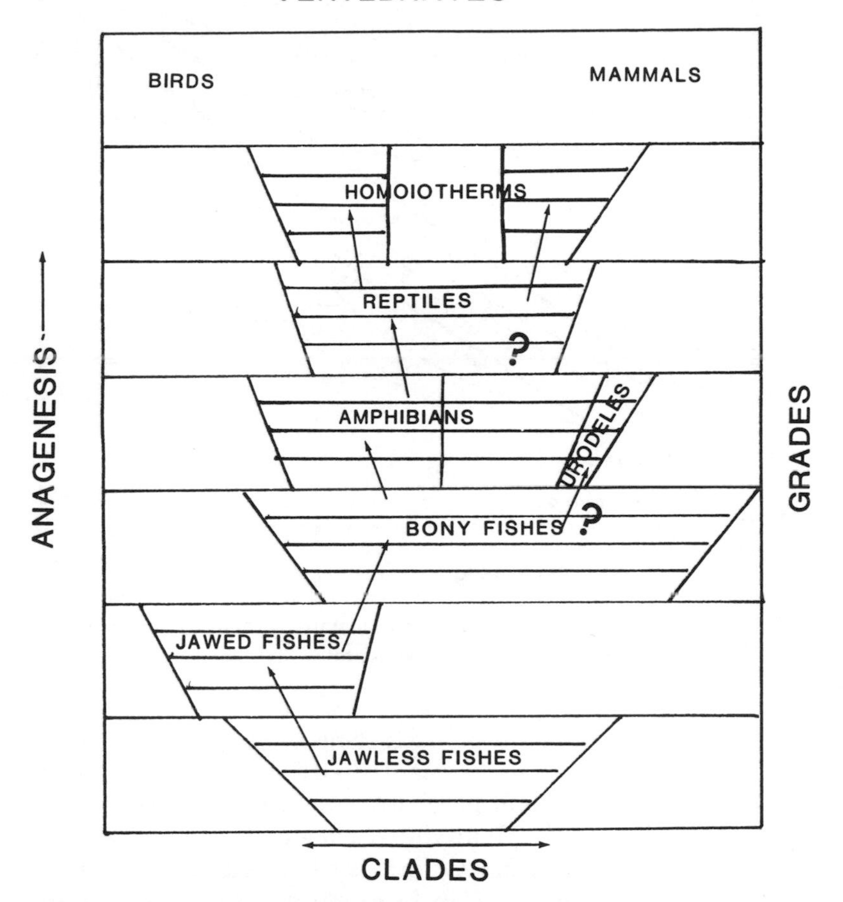

Fig. 14–5. Illustration of anagenesis, grades, and clades in vertebrate evolution. (Adapted from Julian Huxley, 1957.)

clade because of the difference in their respective reptilian ancestors (Figure 14–4), they are in the same "anagenetic" (progressive) grade with respect to their similar (although separately evolved) increase in brain:body ratio (to be described in more detail subsequently). Illustrations of the concepts of anagenesis (progressive evolution), clade, and grade in vertebrate evolution are shown in Figure 14–5.

After J. Huxley, Bernhard Rensch (1959), and other evolution-

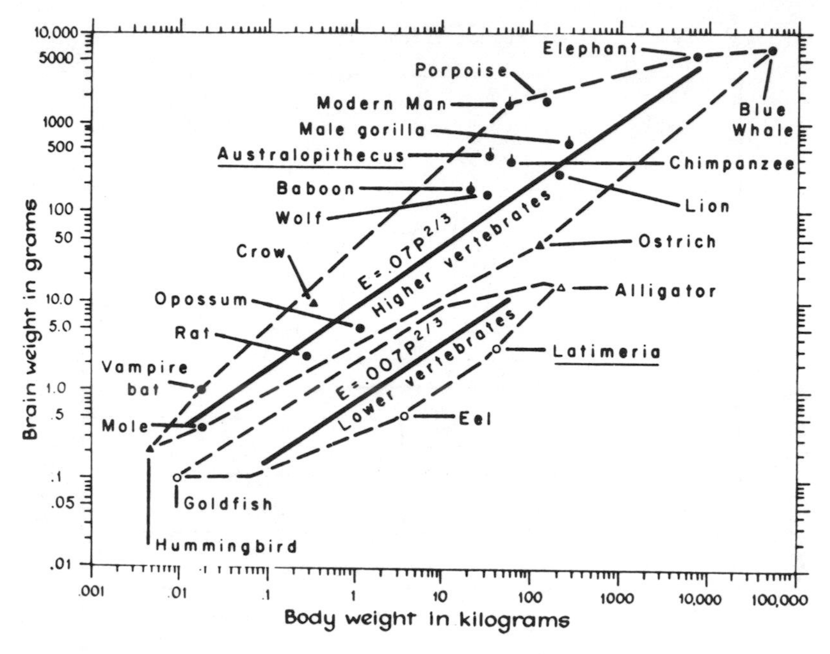

Fig. 14–6. Brain:body ratios of birds and mammals ("higher vertebrates") vs. fish, amphibians, reptiles ("lower vertebrates"). (From Jerison, 1969.)

ary biologists agreed on the pinnacle status of birds and mammals with respect to anagenesis, primarily based on considerations of ontogenetic and behavioral plasticity, a comparative psychologist named Harry Jerison (1973) produced a monumental tome, *Evolution of the Brain and Intelligence,* in which he was able to show that birds and mammals are in a class by themselves as far as the evolution of brain:body ratio is concerned. As can be seen in Figure 14–6, at any given body weight, birds and mammals have a higher brain weight than all species of lower vertebrate at the same body weight. Consequently, according to the ideas already developed in this chapter, we should expect to see, by and large, greater behavioral plasticity in birds and mammals than in lower vertebrates. As it happens, that prediction does accord rather well with learning ability, conceived of as a species' ability to show forms of learning above the level of conditioning: "sensory preconditioning and learned stimulus configuring is [*sic*] possible only in birds and mammals"

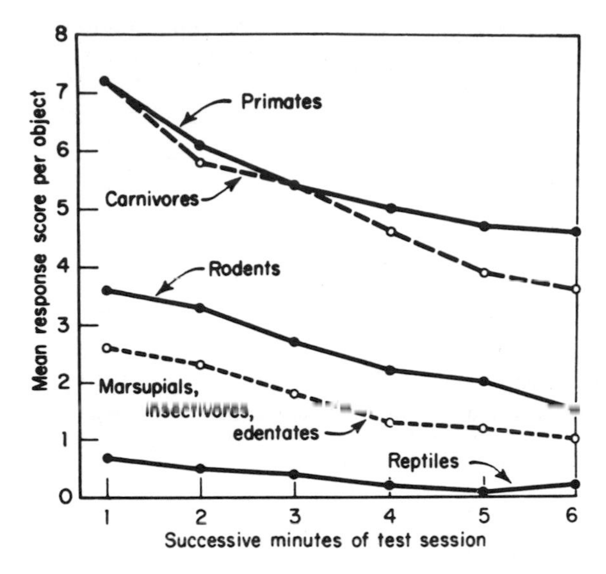

Fig. 14–7. The number of exploratory approaches made to novel objects by a large number of mammalian and reptilian species. (From Glickman & Sroges, 1966.)

(Razran, 1971). From the present standpoint, one of the most interesting findings in an ambitious experimental study of exploratory behavior in a very large variety of vertebrates is the clear superiority of all mammalian forms tested (Figure 14–7). In their study, Glickman and Sroges (1966) studied over 300 animals in over 100 different species, certainly the most extensive survey of exploratory behavior ever undertaken, so these results are most impressive from the standpoint of the consistency of mammalian superiority over the other vertebrate species.

It has been our contention that exploratory behavior—when a species is sufficiently plastic to initiate it—places the individual in a different niche facing different selective (adaptive) demands and thereby brings out latent morphological changes that then allow a genetically based evolutionary change to follow in its wake, as summarized in Table 14–1.

This sort of scheme allows evolution to proceed at a much more rapid rate than is the case when a species must await a severe environmental change or catastrophe such as oceans drying up or

TABLE 14–7. Brain Size in Relation to Rate of Anatomical Evolution

Taxonomic Group	*Relative Brain Size*	*Anatomical Rate*
Homo	114	>10
Hominoids	26	2.5
Songbirds	23	1.6
Other mammals	12	0.7
Other Birds	4.3	0.7
Lizards	1.2	0.25
Frogs	0.9	0.23
Salamanders	0.8	0.26

From Wyles, Kunkel, & Wilson (1983).

months of darkness to prune individuals that are capable of adapting to the change from those that are not. Such extreme environmental changes are rare and widely spaced in time, compared to the frequency and tempo of evolution that could be instigated by behavioral neophenotypes. Consequently, given the relationship between large brains and behavioral plasticity, behavioral neophenogenesis predicts that species with large brains should show evidence of a faster evolutionary pace than species with smaller brains. That prediction accords rather well with the finding of Wyles, Kunkel, and Wilson (1983) of an almost perfect correlation between relative brain size and rate of anatomical evolution for a large number of vertebrate species.

As shown in Table 14–7, humans, the group with the largest relative brain size (Figure 14–7), show the fastest rate of anatomical evolution, with the larger and older hominoid groups ranking second in brain size and rate of evolution. What is perhaps most interesting is that the relatively recently evolved songbirds rank just below hominoids and well above other mammals and other birds on both relative brain size and rate of anatomical evolution. Finally, the classifications of "Other mammals" and "Other birds" show a relatively larger brain size than the lower vertebrates (lizards, frogs, salamanders) and a corresponding faster tempo of evolutionary change. Consonant with the present theory, but without the developmental component, Wyles and colleagues invoke behavioral innovation, as well as a large gene pool, as the major driving force for the observed differences in rates of anatomical evolution: "Behavioral innovation refers to the nongenetic (or genetic) origin of a new skill in a particular individual, leading it to exploit the environment in a new way . . . [the] nongenetic propagation of new skills and mobility in large populations will accelerate anatomical evolution by increasing the rate at which anatomical mutants of potentially high fitness are exposed to selection in new contexts" (Wyles et al., 1983, p. 4396).

Stage II: Change in Morphology without Change in Genes

The present viewpoint takes advantage of the well-accepted fact that only a very small proportion of an individual's genotype participates in the developmental process. Thus, behavioral and morphological phenotypic changes can be immediately instigated by a change in an individual's developmental conditions. In our view, a change in developmental conditions activates heretofore quiescent genes, thus changing the usual developmental process and resulting in an altered behavioral or morphological phenotype. Consequently, stage II in the evolutionary pathway (Table 14–1) holds that the new environmental relationships bring out latent possibilities for morphological–physiological change in advance of the usual criterion of evolution: a change in structural genes or gene frequencies in the population (stage III). It is very exciting that this course of events corresponds to what is known about the correlation of morphological and genetic change in the evolutionary record. According to the best measurement techniques currently available, it would appear that morphological change antedates structural genetic and chromosomal change among the major vertebrate groups (Larson, Prager, & Wilson, 1984; Sarich, 1980, and references therein). Such a state of affairs is precisely compatible with the notion that behavioral neophenogenesis initiates a process of anatomical change that culminates only later in genic and chromosomal change.[3] Since bringing in extragenetic or supragenetic considerations into the evolutionary process, even in its introductory phases, is unorthodox, I would like to point out that this view can be integrated with population-genetic thinking and the modern synthesis (see next section), if the reality or plausibility of the early supragenetic stages I and II can be granted. Since those working on the problem of the apparent lack of synchrony in morphological and chromosomal evolution have themselves looked to behavior (specifically, social behavior) as the means whereby morphological change may be accelerated over chromosomal change in the course of evolution (Larson et al., 1984), the present author's contribution may be seen in the addition of the developmental dimension and its influence on the larger category of behavioral plasticity (exploratory behavior, problem-solving) that is so essential to the early supragenetic phases of evolution. It is now acknowledged in many different quarters, both within and without the modern synthesis (e.g., Bateson, 1988; Futuyma, 1988; Goodwin, 1984; Ho & Saunders, 1982; Johnston & Gottlieb, 1990; Løvtrup, 1987; Oyama, 1985; Rosen & Buth, 1980; Thomas, 1971), that the time has come to include the role of individual development in evolution and I cannot but hope that this small book makes some contribution toward that aim.

Integration of Individual Development into the Population-Genetic Model

Finally, to more explicitly integrate the present theory into the modern synthesis, the evolutionary pathway described here (Table 14–1) is consistent with the idea championed by Parsons (1981), among others, that a behaviorally mediated ecological independence precedes reproductive isolation when populations within a species first begin to split off (speciate). That is, these authors see behaviorally mediated changes in habitat selection (stage II in my scheme), especially microhabitat preferences, as the first step in the pathway to eventual speciation (reproductive isolation). According to this view, when reproductive isolation eventually occurs, it is based on a later developing genetic incompatibility (stage III here) between originally homogeneous gene pools. According to the present scenario, the cause of the original behavioral divergence (e.g., preferences for different temperatures, humidities, light intensities, diets, oviposition sites, and mating sites in fruitflies) would be found in differences in the developmental histories (stage I) of the individuals showing the divergent behaviors. The stage I or developmental contribution to behavioral differences mediating speciation events has not yet been widely appreciated in the literature of evolutionary biology, even where novel shifts in behavior are seen as the key to speciation. That is to say, the causes of the novel shifts in behavior have received little attention as yet (Parsons, 1981, p. 230). The present scenario, specifically the induction of a behavioral neophenotype through a change in developmental conditions, offers an explanation for the novel behavior.

In summary, whereas the architects of the modern synthesis all agree that the mechanisms of evolution are mutation or genetic recombination, selection, migration, and eventual reproductive isolation, the present work describes how migration (invasion of new habitats or niches) may occur without mutation (or recombination) and selection first initiating a change in genes or gene frequencies. Both laboratory and field research indicate that reproductive isolation (i.e., incipient speciation) can occur without major genetic alterations (reviews in Bush, 1973; Singh, 1989, pp. 445–446). The usual scenario—theoretically speaking—is that a major genetic alteration is the necessary *first* step in speciation (Mayr, 1954). Bush's review indicates that behavior can lead the way to speciation with genetic changes coming into play only later on, which is consonant with the evolutionary model presented here. In this respect, it is noteworthy that Mayr (1988, Essay 28, pp. 541–542) now advocates a pluralistic view of the factors that instigate evolutionary change, one that might not entirely reject the present approach.

Bolder Speculation

A bolder and much more radical proposal would hold out for the possibility of morphological change initiated by behavioral change alone (i.e., without a shift in the physical environment). Tradition is recognized as an important component of animal as well as human behavior (Bonner, 1980), so what would be required under this more radical evolutionary scenario would be a change in some behavioral tradition, especially one affecting the rearing experience of young animals in the process of growing up, so that the subsequent behavior of the developing animals would be altered without necessarily causing them to leave their usual physical environment or niche. So far all the behaviorally mediated evolutionary scenarios have assumed an ecological change. The more radical notion of a strictly behaviorally mediated morphological change leading to speciation without an ecological or environmental shift has not been put forward before, at least to my knowledge. Lamarck's behaviorally mediated morphological evolution was said by him to be stimulated by a physical environmental change or stressor. The controversial Lamarckian experiments by Steele (1979) involved a morphological response to an environmental change. The possibility of strictly behaviorally inspired morphological change in the absence of stage II (Table 14–1) has not been previously entertained. Based on the evidence and other considerations discussed in this monograph, I think it remains a theoretical possibility. For example, a change in behavior almost inevitably brings about a change in social interactions, and some authors theorize that changes in social interaction prompted the evolution of human intelligence (reviewed in Byrne & Whiten, 1988). But these scenarios are couched within the terms of the modern synthesis (genetic change brought about by natural selection leads to evolutionary change in intelligence). In the framework developed here, an evolutionary change in brain structure and function might well occur without a genetic change in the population. There is such a tremendous amount of currently unexpressed developmental potential that behavioral and anatomical evolution would seem possible without the necessity of new genetic variations produced by mutation and genetic recombination.

It is appropriate to ask how the new phenotypic changes can be preserved from one generation to the next if there has not been a mutation or new genetic recombination. The answer is that the transgenerational stability of new behavioral and morphological phenotypes is preserved by the repetition of the developmental conditions that gave rise to them in the first place. Since genes are a part of the developmental system and cannot make traits by themselves, this same requirement (repetition of developmental conditions) holds for new phenotypes stemming from mutation and genetic recombination, even though that requirement often goes unrecognized or unspecified in the modern synthesis account of evolutionary change.

While I am unable to propose a specific molecular mechanism whereby the organism's new experiences activate previous unactivated DNA (to get the expression of previously inactive genes), the present proposal obviously assumes such a mechanism. For example, something akin to the "homeobox" would fit the bill (Gehring,

1987; Ingham, 1988). Some such mechanism is required for the Bolder Speculation to work. As described in an earlier chapter, the activation of previously inactive genes must be occurring when avian oral epithelial cells grow a "mammalian" tooth.

It seems clear to me that the Bolder Speculation applies to our own species: Earlier, we had suggested that *Homo erectus* could have possibly evolved into *Homo sapiens* through a dramatic change in rearing practices (Gottlieb, Johnston, & Scoville, 1982). It remains to be seen whether the behavioral activation of new anatomical features from a previously unexpressed developmental potential is more widely applicable, or whether the more usual scenario is a behaviorally initiated ecological shift bringing out latent morphological change (stage II in Table 14–1). In either event, a developmentally wrought behavioral change would play an important role in evolution.

Conclusion

The present theory of behavioral neophenogenesis is manifestly a theory of vertebrate evolution, particularly of the higher vertebrates (birds and mammals), where the role of early experience in enhancing brain size, learning ability, exploratory behavior, and resistance to stress has been experimentally demonstrated in a number of species. The impact of early experience on later behavior is not without import in invertebrates (e.g., Jaisson, 1975; McDonald & Topoff, 1985) and lower vertebrates, but such developmental studies are few in number, so the question remains more open on the role of behavioral neophenogenesis in the evolution of invertebrate and lower vertebrate forms (fish, amphibians, reptiles). However, our general concept of neophenogenesis (Johnston & Gottlieb, 1990), which is a more global developmental theory of phenotypic evolution, would seem to have broad application in the evolutionary arena.

NOTES

Chapter 1

1. A comprehensive scholarly review of the concepts of epigenesis and preformation appears in Joseph Needham's (1959) *A History of Embryology,* which served as my primary source for seventeenth and eighteenth century views on individual development.

Chapter 2

1. Lamarck is here pointing out the necessity of selective breeding if the acquired modification is to be perpetuated in subsequent generations. In the 1950s, Conrad Waddington was to take the necessary experimental steps to show that individuals that have been modified in the same way by an unusual environmental occurrence early in development, when selectively bred, indeed do "pass on" the alteration in a most unexpected way: the morphological modification eventually becomes "genetically assimilated" so that the original environmental stimulus is no longer required for its production (reviewed in Chapter 11)!

Since Lamarck did not know about genes, much less genetic assimilation, he did not say anything to imply the assimilation effect discovered by Waddington. The interesting point is that, in Lamarck's version of heredity, the perpetuation of acquired characters is achieved by selective mating, which, as seen in the early stages of Waddington's experiments before "genetic assimilation" occurs, is correct (i.e., an increasing number of individuals with the new phenotype across selectively bred generations).

Chapter 3

1. There is a veritable Darwin industry practiced by philosophers and historians of science on Charles Darwin's published and unpublished writings, and I do not pretend to be making any original contribution to that

body of work. A deeply scholarly contribution to the Darwin industry that describes Charles's intellectual struggle with the various possible mechanisms for evolutionary change (and the objections thereto) is R. J. Richards's (1987) *Darwin and the Emergence of Evolutionary Theories of Mind and Behavior.*

2. As further information has become available on the brain:body sizes of more reptilian species, it has become evident that *Archaeopteryx's* brain size (relative to its body size) is considerably larger than Jerison's original estimate, so that now *Archaeopteryx's* relative brain size would be grouped with extant reptiles of its estimated body size (Hopson, 1977). Consequently, Figure 3–3 should be considered only as an illustration of a transitional form (should one exist) between ancient reptiles and birds, at least with respect to brain size. Obviously, from a more comprehensive morphological point of view, *Archaeopteryx* does fit in a transitional zone between reptiles and birds. The whole natural endocast of the London *Archaeopteryx* has been broken out of the limestone slab in which it was embedded, so that a more accurate description of its braincase and its evolutionary lineage has been made possible (Whetstone, 1983).

Chapter 5

1. Mivart uses the term *genetic affinity* several times in the *Genesis* and it is rather frustrating not to be able to ferret out precisely what he had in mind with this phrase some thirty years before the words *gene* and *genetics* were coined to mean about what they do today. Most likely he meant close affinity in origin, without, of course, having genes in mind, but also certainly signifying a hereditary relatedness, the mechanism of which was not yet understood. Hereditary relatedness was understood at the time as "like begetting like" by some unknown means of hereditary transmission. In the twentieth century, psychologists used (and some still use) the term genetic interchangeably with the word developmental—this would also be consistent for Mivart, who placed some emphasis on the importance of development for our understanding of evolution and phylogeny. It is ironic that genetic came to be used in a nondevelopmental way with the rediscovery of Mendel's work in 1900 (discussed in Chapter 8).

Chapter 6

1. These chapters in intellectual history have been reviewed in splendid detail by Daniel J. Kevles (1985) in his *In the Name of Eugenics: Genetics and the Uses of Human Heredity.*

2. The significant implications of the difference between the narrowly circumscribed notion of a reaction range versus the more open and indefinitely bounded notion of a *norm of reaction* are spelled out in detail in Chapter 14.

Chapter 7

1. Weismann's theory of heredity derived from Darwin's theory of pangenesis as modified by Galton (1883/1970). The basic idea was that the germ plasm consisted of a very large number of germs, each corresponding to a unit of the body. Only some of these germs were used up in the formation of the body—the remainder were available for transmission to the next generation. Galton did not accept the idea that these transmission germs could somehow be modified by the experience of the organism to affect the transmission of acquired characters as did Darwin. Weismann's concept of the continuity of the germ plasm was an extension of Galton's notion of the unmodifiability of the germ, thus seeing heredity as the continuity of the germ plasm.

Chapter 10

1. In the West it is usual to recognize R. A. Fisher, J. B. S. Haldane, and Sewall Wright as the principal founders of the field of population genetics. They laid the necessary statistical and mathematical groundwork for what was later to be called the modern synthesis (Chapter 11). However, a little-known Russian biologist, Sergei Sergeevich Chetverikov (also spelled Tshetverikov), is recognized by Dobzhansky (1970) as playing an important role in the establishment of population genetics in the East. Unfortunately, Chetverikov published relatively little, and only one of his articles (1926/1961) has been translated into English, so his contribution seems fated to remain little appreciated in the West, a regrettable state of affairs to which this meager footnote bears testament.

2. In his recent biography of Sewall Wright, William Provine concludes that Wright will ultimately be seen "as perhaps the single most influential evolutionary theorist of this century" (1986, p. 499). It is interesting to speculate that Wright's failure to be so appreciated now is related to his unconventional views (among population thinkers) on the intimate relationship of development to evolution: "At all times Wright saw his work in physiological [i.e., developmental] genetics as fundamentally related to and fully integrated with his ideas about the evolutionary process" (Provine, 1986, p. 205).

3. Platt and Sanislow (1988) have written an important critique of the inappropriate use and misunderstanding of the concept of genetic limitations. The fact that the authors are specialists in behavioral genetics lends special authority to the critique.

Chapter 11

1. It is usual to include among the architects of the modern synthesis, in addition to the persons cited more extensively in Chapter 11, George

Gaylord Simpson (1944), who discriminated the various tempos and modes of evolution, Bernhard Rensch (1959), who dealt particularly with evolution above the species level, and G. Ledyard Stebbins, Jr. (1950), the only member of the group to cover botany. In the introduction to the 1984 reissue of his 1944 book, G. G. Simpson is somewhat less generous in that he recognizes only himself, Dobzhansky, Mayr, and Stebbins as the principal architects of the modern synthesis.

2. While Schmalhausen's term *stabilizing selection* has been incorporated into the modern synthesis, Schmalhausen's concept of an eventual transgenerational change in developmental dynamics (also see Footnote 3) has been completely overlooked. The more static modern synthesis notion of stabilizing selection is a transgenerational narrowing of the range of phenotypic expression by selective mating and enhanced reproductive success around the mean of a given phenotypic feature, thereby reducing the extreme expressions on both sides of the mean, enhancing the mean value, and thus reducing variability (Lerner, 1958, p. 7, Fig. 1.1; Trivers, 1985, pp. 22–23; Figure 11–6, this volume). In that usage of Schmalhausen's term, the phenotype under selection is an adaptation rather than an adaptability. Schmalhausen's point about adaptability being an insufficiently studied aspect of evolutionary theory still holds today. In the writer's opinion, that is primarily due to the erroneous notion that adaptations are genetically based, whereas adaptabilities are not (Mayr's dichotomy on sources of variation). Theorists that see adaptability as having genetic correlates and thus treating it as an essential part of evolution would seem to be in the minority (e.g., Conrad, 1983). In Chapter 14 of this volume I try to show the significance of behavioral adaptability for evolutionary change, adopting the point of view that there must be genetic correlates to behavioral or any other kind of adaptability.

3. I would be remiss if I did not record what I feel to be a small weakness in Schmalhausen's highly insightful, virtually forgotten conception of stabilizing selection II. As he makes clear in the preface to the English translation of his book, Schmalhausen sees stabilizing selection II as a three-step process. First, the new morphological adaptation is brought about by a change in external circumstances (i.e., it is stimulated or "caused" by environmental factors). Next, the adaptation comes to be controlled by internal events that exert canalizing influences on its ontogenetic development. In the third and final stage, Schmalhausen asserts that the new adaptation arises via "autonomous development." The second stage he sees as the same as Waddington's notion of developmental canalization, which the latter put forward to try to understand the process of the so-called genetic assimilation of an originally environmentally induced character.

Although Schmalhausen identifies the third stage as his distinctive theoretical contribution, I see no way that one could truly distinguish between the second and third stages in that the new adaptation appears in both of them without the necessity of the original environmental stimulating circumstance that initially gave rise to the morphological change. It is difficult to imagine a state of affairs that would be any more canalized than the second stage, but in the preface Schmalhausen (1949, p. VIII) says the third stage reflects autonomous development. In the part of the book where he discusses the "Autonomization of Ontogeny" (pp. 237–243), he talks of the trend toward increased internal canalization during the process of evolution, so it is not possible to distinguish a second and third stage except by thinking of the third stage as a somehow more deeply or further canalized stage of stage two. In other words, I don't think there is anything clearly qualitatively different about stages two and three, although Schmalhausen obviously believed differently.

4. In fairness to Lamarck, few people notice that he made the inheritance of an acquired characteristic contingent on both members of the breeding pair exhibiting the characteristic in question (Chapter 2). Lamarck did not, of course, predict the unanticipated outcome of Waddington's experiments, but his statements are consistent with the increase in the selected phenotype in the early generations before the eventual, so-called genetic assimilation of the acquired character.

5. In their curious definition of genetic, population geneticists would want to say that the lack of heritable variation is "nongenetic" in the sense that the lack of differences implies the absence of genetic differences = nongenetic. But they also believe genes are correlated with traits, so it does get confusing for everyone, as indicated in the previous chapter.

6. G. C. Williams (1982) uses another of Waddington's (1956) experiments—induction of the bithorax condition by ether—to discredit the notion of genetic assimilation as a possible factor in evolution (but see his postscript on p. 202 of that chapter). Williams's conclusion is that the ether ("new environment") that induces the bithorax phenotype should be viewed as a toxin that "prevented the normal developmental plan from producing a normal phenotype" [p. 203], and that the bithorax flies are degenerate forms that shed no light on evolutionary mechanisms. Bithorax flies have two sets of wings instead of one set. The second set of wings develops in place of the halteres (balancing organs), so the bithorax phenotype is maladaptive in that it cannot balance itself in order to feed or otherwise behave in an adaptive fashion that would promote its individual survival. Whether the bithorax is instructive or not with respect to evolution may depend on the theoretical framework in which the experiments are carried out (see, e.g., Ho, 1983, 1984). In the text I have used another of Waddington's genetic assimilation experiments involving the development of enlarged anal papillae in response to excess salt as reflecting a positive adaptation, thereby sidestepping Williams's criticism of degeneracy with reference to the bithorax experiments.

Chapter 12

1. In 1918, Valentin Haecker published a book called *Entwicklungsgeschichtliche Eigenschaftsanalyse* (*Phänogenetik*), which was intended to launch a new scientific approach to the problem of understanding differences between closely related species by comparing their ontogenetic developmental pathways for the particular phenotypic characters in question. As Haecker saw it, phenogenetics, the development analysis of characters, differed from developmental mechanics and developmental physiology in that it focuses on the mature characteristics of related species and traces these back to their developmental origin, i.e., to their first bifurcation in development or "phenocrisis." (The traditional way of embryological study begins with the fertilized egg and proceeds in a forward direction.) Haecker's idea was that developmental differences during the phenocrisis would explain differences in the mature character in the species or breeds under comparison. Although Haecker envisioned his phenogenetics as an effort to integrate development, heredity, and evolution, it failed to synthesize those fields, which, as the previous chapters reveal, were each steadfastly pursuing their own independent directions at the time.

While the present use of the term phenogenesis obviously owes something to Haecker's concept, the current aim is to provide a general orienting framework for the study of individual development, especially as such has ramifications for our understanding of evolution.

I would like to acknowledge the generous assistance of Dr. Elga Wulfert, who translated a number of particularly difficult passages that helped to clarify Haecker's central ideas.

2. Ho's work is an extension of that of Viktor Jollos (1934), whose terminal career has been described in moving detail by Saap (1987). While it was not Jollos's cytoplasmic inheritance experiments alone that brought his career and his life to such a sorrowful end, it is clear from Saap's account that those unconventional experiments were an essential component of Jollos's demise. That the intellectual climate has not changed much in the ensuing years is indicated by Mae-Wan Ho's difficulty in securing an outlet for her findings (M.-W. Ho, personal communication, 1987). It was not the soundness of Ho's experiments that was at issue—it was that the editors of various journals would not want to publish her results even if they were sound!

3. The outcome of Tryon's selection experiment, as in other selective breeding experiments, was highly specific to the conditions of rearing and testing. When the "bright" strain was compared to the "dull" strain on other learning and performance tasks, the dull strain was sometimes superior and often equal to the bright strain (Searle, 1949). Thus the brightness factor did not extend across all or even many of the learning tasks. According to Searle's results there is no general intelligence factor and the results of

selective breeding are highly specific to the conditions under which the selection is carried out. This important qualification will be documented further in Chapter 14, where the concepts of reaction range and norm of reaction are discussed.

4. Professor Lewontin now feels it is more correct to say *almost* any trait can be selected for in *Drosophila* (personal communication, October 23, 1989).

5. Not all genes are expressed during individual development; in fact, only a small minority are expressed. The term *genetic store* is intended to call attention to the fact that, during development, different gene complexes can be active in otherwise isogenic (or closely related) organisms, thus contributing to eventual differences in the phenotype. An alternative—not mutually exclusive—is that different supragenetic influences (e.g., cytoplasmic differences) alter the expression of the same gene complexes, thereby contributing to eventual phenotypic differences. That is, the same germ complexes may not necessarily produce the same specific proteins each time they are activated. The DNA–RNA–protein relationship itself is influenced by supragenetic factors (see Lewin, 1980; Pritchard, 1986). Thus, there may not be a "built-in" fidelity between specific DNA complexes and specific proteins. In this latter case, it would be more appropriate to think of the Enormous Hidden Developmental Store for Variation, rather than a strictly Genetic Store. Although necessarily speculative, these two not-mutually-exclusive alternatives are in line with the evidence and point of view presented elsewhere in the text.

Chapter 13

1. In 1976 I (Gottlieb, 1976a, b) defined experience in such a way as to include spontaneous activity generated within the nervous system as well as evoked activity arising from sensory stimulation originating in the organism's environment. Since that time it has been shown that spontaneous as well as evoked activity does play an important role in the normal neuroanatomical development of the brain (e.g., Born & Rubel, 1988; Shatz & Stryker, 1988). The brain develops abnormally (deficiently) when normal spontaneous or evoked activity is curtailed by experimental means. This area of research is coming to the forefront in developmental neuroscience under the rubric *activity-dependent regulation of gene expression,* a concept that meshes extremely well with the view of the mechanisms of individual development expressed in the present work (review in Changeux & Konishi, 1987). The contribution of factors "upstream" from the genes is also being recognized by the "new look" in developmental neuroscience (e.g., Edelman, 1988).

2. With the realization that Weismann's notion that specific genes give rise to specific parts of the body is erroneous, various biologists have been

working on hierarchically organized field-theory approaches to an understanding of individual morphological development. Among the current workers are Goodwin (1984), Hall (1988), and Oster, Odell, & Alberch, (1980). It is curious that some ostensible field theorists such as Wolpert (1971, 1982) still hold that the morphological development of the individual is somehow preprogrammed in the genes. So, when I say "the realization that Weismann's view is erroneous," it is obvious that the implications of that statement are not unequivocally understood by all developmental biologists, even if they happen to be working on field theories of individual development!

3. It is especially noteworthy that someone like Futuyma acknowledges the lack of developmental thinking in the neo-Darwinian concept of evolution, because he is firmly identified with the population-genetic tradition that undergirds the modern synthesis. On the other hand, Futuyma seems to see no substantive difference or change in the neo-Darwinian synthesis resulting from an inclusion of developmental considerations: "The need for a theory of development . . . is evident, but at this point, the need for a new evolutionary paradigm is not" (Futuyma, 1988, p. 221). Other biologists do not agree with Futuyma; they try to show why a new evolutionary paradigm *is* necessary when individual development is taken into account (Ho & Saunders, 1984; Ho & Fox, 1988). In Chapter 14 I try to show what differences accrue to evolutionary thinking when developmental behavioral and psychological concerns are placed in the forefront.

Chapter 14

1. The following review is highly selective. A large number of laboratories contributed to the explosion of developmental-experiential studies on brain and behavior in the 1950s and 1960s. Among the most prominent was the Berkeley group of David Krech, Mark Rosenzweig, Edward Bennett, and their various students and collaborators. That line of research inspired considerable research all over the world (summarized in Renner & Rosenzweig, 1987). An important point demonstrated by the "environmental enrichment" approach is that the brain remains plastic throughout life. However, the plasticity of the brain in the young animal is greater than that of adult animals (studies cited in Renner & Rosenzweig, p. 72), which is basic to the present point of view. Also, Renner has taken the behavioral analysis to a new level and discovered that it is not the quantity but the quality of the interaction with novel objects that distinguishes the adult exploratory behavior of the early enriched from early isolated animals (summarized in Renner & Rosenzweig), which is a significant contribution and highly compatible with the view advocated here.

2. I have not reviewed the small literature on the results of early enrichment on learning and brain development in birds. The few behavioral

studies that have been done (e.g., Heaton & Klein, 1981) are entirely in keeping with the positive effects on later problem-solving observed in mammals. Also, with respect to brain development, it is clear that it is not merely the size of the brain that counts in large-brained species but the areas of prominence (e.g., *wulst*) that are underdeveloped in species with lesser learning capabilities (Krushinsky, 1965). The avian brain is sensitive to early experiential alterations (Horn, 1985; Rogers, 1990) but has been much less studied in that respect than the mammalian brain. Certain species of songbirds retain an amazing degree of cerebral and behavioral plasticity into adulthood (Alvarez-Buylla, Kirn, & Nottebohm, 1990).

3. Allan Wilson (1985) holds that the exceptionally large brains of animals such as mammals and birds endow them with an internal pressure to evolve. Although Wilson's scheme lacks an explicit developmental dimension, large brains and altered behavior play key roles in his view of evolutionary change, particularly the rate of evolutionary change. I think that otherwise the present scenario and Wilson's have much in common. It was he and his coworkers, Jeff S. Wyles and Joseph G. Kunkel, who first pointed out the correlation between brain size and evolutionary rate (Wyles, Kunkel, & Wilson, 1983).

REFERENCES

Alberch, P. (1980). Ontogenesis and morphological diversification. *American Zoologist, 20,* 653–667.

Alvarez-Buylla, A., Kirn, J. R., & Nottebohm, F. (1990). Birth of projection neurons in adult avian brain may be related to perceptual or motor learning. *Science, 249,* 1444–1446.

von Baer, K. E. (1828). *Ueber Entwickelungsgeschichte der Thiere: Beobachtung und Reflexion. Part one.* Königsberg: Bornträger. (Reprinted 1966 by Johnson Reprint Corporation.)

von Baer, K. E. (1828/1853). Philosophical fragments. Translated (1853) in A. Henfrey & T. H. Huxley (Eds.), *Scientific memoirs, selected from the transactions of foreign academies of science and from foreign journals. Natural History.* Vol. 1, part 3. London: Taylor and Francis.

Baldwin, J. M. (1902). *Development and evolution.* New York: Macmillan.

Barlow, N. (Ed.) (1958). *The autobiography of Charles Darwin 1809–1882.* New York: Harcourt, Brace.

Bateson, B. (1928). *William Bateson, F.R.S., naturalist.* Cambridge: Cambridge University Press.

Bateson, P. (1983). Genes, environment, and the development of behaviour. In T. R. Halliday & P. J. B. Slater (Eds.), *Animal behaviour,* vol. 3. *Genes, development, and learning.* Oxford: Blackwell.

Bateson, P. (1988). The active role of behaviour in evolution. In M.-W. Ho & S. W. Fox (Eds.), *Evolutionary processes and metaphors.* London: Wiley.

Bateson, W. (1894). *Materials for the study of variation, treated with especial regard to discontinuity in the origin of species.* London and New York: Macmillan.

Bateson, W. (1913). *Problems of genetics.* New Haven: Yale University Press. (Reprinted in 1979).

de Beer, G. (1930). *Embryology and evolution.* Oxford: Clarendon Press.

de Beer, G. (1954). "*Archaeopteryx lithographica.*" British Museum of Natural History, London.

de Beer, G. (1958). *Embryos and ancestors,* third ed. Oxford: Clarendon Press.

von Bertalanffy, L. (1962). *Modern theories of development: An introduction to theoretical biology.* New York: Harper. (Originally published in German in 1933.)

Bolk, L. (1926). *Das Problem der Menschwerdung.* Jena: Gustav Fischer.

Bonner, J. T. (1980). *The evolution of culture in animals.* Princeton, New Jersey: Princeton University Press.

Bonner, J. T. (1983). How behavior came to affect the evolution of body shape. *Scientia, 118,* 175–183.

Born, D. E., & Rubel, E. W. (1988). Afferent influences on brain stem auditory nuclei of the chicken: Presynaptic action potentials regulate protein synthesis in nucleus magnocellularis neurons. *Journal of Neuroscience, 8,* 901–919.

Broadhurst, P. L. (1965). The inheritance of behavior. *Science Journal* (London), *24,* 39–43.

Bronfenbrenner, U. (1979). *The ecology of human development: Experiments by nature and design.* Cambridge, Massachusetts: Harvard University Press.

Brooks, W. K. (1883). *The law of heredity,* 2nd ed. revised. Baltimore: John Murphy.

Brooks, W. K. (1902). The intellectual conditions for embryological science. II. *Science, 15,* 481–492.

Burkhardt, R. W., Jr. (1977). *The spirit of system: Lamarck and evolutionary biology.* Cambridge, Massachusetts: Harvard University Press.

Bush, G. L. (1973). The mechanism of sympataric host race formation in the true fruit flies (*Tephritidae*). In M. J. D. White (Ed.), *Genetic mechanisms of speciation in insects.* Dordrecht, Holland: D. Reidel.

Byrne, R. W., & Whiten, A. (Eds.) (1988). *Machiavellian intelligence: Social expertise and the evolution of intellect in monkeys, apes, and humans.* Oxford: Oxford University Press.

Cairns, R. B. (1979). *Social development: The origins and plasticity of interchanges.* San Francisco: W. H. Freeman.

Cavalier-Smith, T. (1985). Cell volume and the evolution of eukaryote genome size. In T. Cavalier-Smith (Ed.),*The evolution of genome size.* Chichester, England: Wiley.

Changeux, J.-P., & Konishi, M. (Eds.) (1987). *The neural and molecular bases of learning.* Chichester, England: Wiley.

Cheng, M.-F. (1979). Progress and prospects in ring dove: A personal view. *Advances in the Study of Behavior, 9,* 97–129.

Chetverikov, S. S. (1961). On certain aspects of the evolutionary process from the standpoint of modern genetics. *Proceedings of the American Philosophical Society, 105,* 167–195. (Originally published in Russian in 1926.)

Cierpal, M. A., & McCarty, R. (1987). Hypertension in SHR rats: Contribution of maternal environment. *American Journal of Physiology, 253,* 980–984.

Clark, N. M., & Galef, B. G. (1988). Effects of uterine position on rate of sexual development in female mongolian gerbils. *Physiology & Behavior, 42,* 15–18.

Conrad, M. (1983). *Adaptability: The significance of variability from molecule to ecosystem.* New York: Plenum Press.

Cooper, R. M., & Zubek, J. P. (1958). Effects of enriched and restricted early environments on the learning ability of bright and dull rats. *Canadian Journal of Psychology, 12,* 159–164.

Darwin, C. (1859). *On the origin of species.* (A facsimile of the 1st edition, published by Harvard University Press in 1964.)

Darwin, E. (1794). *Zoonomia; or the laws of organic life.* London: Johnson.

Davidson, E. H. (1986). *Gene activity in early development.* Orlando, Florida: Academic Press.

Dawson, W. M. (1932). Inheritance of wildness and tameness in mice. *Genetics, 17,* 296–326.

Denenberg, V. H. (1964). Critical periods, stimulus input, and emotional reactivity: A theory of infantile stimulation. *Psychological Review, 71,* 335–351.

Denenberg, V. H. (1969). The effects of early experience. In E. S. E. Hafez (Ed.), *The behaviour of domestic animals,* 2nd ed. Baltimore: Williams & Wilkins.

Denenberg, V. H., & Rosenberg, K. M. (1967). Nongenetic transmission of information. *Nature, 216,* 549–550.

Desmond, A. (1982). *Archetypes and ancestors. Paleontology in Victorian England 1850–1875.* Chicago: University of Chicago Press.

Dewey, J., & Bentley, A. F. (1949). *Knowing and the known.* Boston: Beacon.

Dewsbury, D. A. (1978). *Comparative animal behavior.* New York: McGraw-Hill.

DiBerardino, M. A. (1988). Genomic multipotentiality of differentiated somatic cells. In G. Eguchi, T. S. Okada, Saxén (Eds.), *Regulatory mechanisms in developmental processes.* Ireland: Elsevier.

Dobzhansky, T. (1951). *Genetics and the origin of species,* 3rd ed. New York: Columbia University Press.

Dobzhansky, T. (1955). *Evolution, genetics, and man.* New York: Wiley.

Dobzhansky, T. (1970). *Genetics of the evolutionary process*. New York: Columbia University Press.

Driesch, H. (1908/1929). *The science and philosophy of the organism*. London: A. & C. Black. (The 2nd abridged edition was used in the present work.)

Dunn, L. C. (1965). *Short history of genetics*. New York: McGraw-Hill.

Edelman, G. M. (1988). *Topobiology*. New York: Basic Books.

Eisenberg, L. (1976). The outcome as cause: Predestination and human cloning. *Journal of Medicine and Philosophy, 1,* 318–331.

Emerson, A. E. (1958). The evolution of behavior in social insects. In A. E. Roe, & G. G. Simpson (Eds.), *Behavior and evolution*. New Haven: Yale University Press.

Ephrussi, B. (1979). Mendelism and the new genetics. *Somatic Cell Genetics, 5,* 681–695.

Ewing, A. W. (1961). Body size and courtship behaviour in *Drosophila melanogaster. Animal Behaviour, 9,* 93–99.

Feldman, M. W., & Lewontin, R. C. (1975). The heritability hang-up. *Science, 190,* 1163–1168.

Fisher, R. A. (1918). The correlation between relatives on the supposition of Mendelian inheritance. *Transactions of the Royal Society of Edinburgh, 52,* 399–433.

Fisher, R. A. (1930). *The genetical theory of natural selection*. Oxford, England: Oxford University Press. (A 2nd revised edition was published by Dover in 1958).

Ford, E. B., & Huxley, J. S. (1927). Mendelian genes and rates of development in *Gammarus chevreuxi. British Journal of Experimental Biology, 5,* 112–134.

Forgays, D. G., & Forgays, J. W. (1952). The nature of the effect of free-environmental experience in the rat. *Journal of Comparative and Physiological Psychology, 45,* 322–328.

Frings, H., & Frings, M. (1953). The production of stocks of albino mice with predictable susceptibilities to audiogenic seizures. *Behaviour, 5,* 305–319.

Futuyma, D. J. (1988). *Sturm und Drang* and the evolutionary synthesis. *Evolution, 42,* 217–226.

Galton, F. (1869). *Hereditary genius: An inquiry into its laws and consequences*. London: Macmillan. (The work referred to in Chapter 8 is the 1892 edition.)

Galton, F. (1875). *English men of science: Their nature and nurture*. New York: D. Appleton. (Originally published in 1874.)

Galton, F. (1907). *Inquiries into human faculty and its development*. London: J. M. Dent and New York: E. P. Dutton. (Originally published in 1883.)

Garstang, W. (1922). The theory of recapitulation: A critical re-statement of the biogenetic law. *Journal of the Linnean Society of London, Zoology, 35,* 81–101.

Gehring, W. J. (1987). Homeo boxes in the study of development. *Science, 236,* 1245–1252.

Gerhart, J. C. (1987). The epigenetic nature of vertebrate development: An interview of Pieter D. Nieukoop on the occasion of his 70th birthday. *Development, 101,* 653–657.

Glickman, S. E., & Sroges, R. W. (1966). Curiosity in zoo animals. *Behaviour, 26,* 151–188.

Goldschmidt, R. (1933). Some aspects of evolution. *Science, 78,* 539–547.

Goldschmidt, R. (1938). *Physiological genetics.* New York: McGraw-Hill.

Goldschmidt, R. (1952). Evolution, as viewed by one geneticist. *American Scientist, 40,* 84–98, 135.

Goodman, C. S. (1978). Isogenic grasshoppers: Genetic variability in the morphology of identified neurons. *Journal of Comparative Neurology, 182,* 681–705.

Goodwin, B. C. (1984). A relational or field theory of reproduction and its evolutionary implications. In M.-W. Ho, & P. T. Saunders (Eds.), *Beyond neo-Darwinism: An introduction to the new evolutionary paradigm.* London: Academic Press.

Gorbman, A., Dickhoff, W. W., Vigna, S. R., Clark, N. B., & Ralph, C. L. (1983). *Comparative endocrinology.* New York: Wiley.

Gottesman, I. I. (1963). Genetic aspects of intelligent behavior. In N. Ellis (Ed.), *The handbook of mental deficiency.* New York: McGraw-Hill.

Gottesman, I. I., & Shields, J. (1982). *Schizophrenia: The epigenetic puzzle.* Cambridge, England: Cambridge University Press.

Gottlieb, G. (1970). Conceptions of prenatal behavior. In L. R. Aronson et al. (Eds.), *Development and evolution of behavior.* San Francisco: W. H. Freeman.

Gottlieb, G. (1976a). The roles of experience in the development of behavior and the nervous system. In G. Gottlieb (Ed.), *Neural and behavioral specificity.* New York: Academic Press.

Gottlieb, G. (1976b). Conceptions of prenatal development: Behavioral embryology. *Psychological Review, 83,* 215–234.

Gottlieb, G. (1985). Development of species identification in ducklings: XI. Embryonic critical period for species-typical perception in the hatchling. *Animal Behaviour, 33,* 225–233.

Gottlieb, G., Johnston, T. D., & Scoville, R. P. (1982). Conceptions of development and the evolution of behavior. *Behavioral and Brain Sciences, 2,* 284.

Gould, S. J. (1977). *Ontogeny and phylogeny.* Cambridge, Massachusetts: Harvard University Press.

Greenough, W. T., & Juraska, J. M. (1979). Experience-induced changes in brain fine structure: Their behavioral implications. In M. E. Hahn, C. Jensen, & B. C. Dudek (Eds.), *Development and evolution of brain size*. New York: Academic Press.

Grene, M. (1987). Hierarchies in biology. *American Scientist, 75*, 504–510.

Grouse, L. D., Schrier, B. K., & Nelson, P. G. (1979). Effect of visual experience on gene expression during the development of stimulus specificity in cat brain. *Experimental Neurology, 64*, 354–359.

Grouse, L. D., Schrier, B. K., Letendre, C. H., & Nelson, P. G. (1980). RNA sequence complexity in central nervous system development and plasticity. *Current Topics in Developmental Biology, 16*, 381–397.

Gruber, J. (1960). *A conscience in conflict*. New York: Columbia University Press.

Guhl, A. M., Craig, J. V., & Mueller, L. D. (1960). Selective breeding for aggressiveness in chickens. *Poultry Science, 39*, 970–980.

Gurdon, J. B. (1968). Transplanted nuclei and cell differentiation. *Scientific American, 219*, 24–35.

Haeckel, E. (1866). *Generelle Morphologie der Organismen: Allgemeine Grundzuge der organischer Formen-Wissenschaft, mechanisch begrundet durch die von Charles Darwin reformierte Descendenz-Theorie*, 2 vols. Berlin: Reimer.

Haeckel, E. (1891). *Anthropogenie oder Entwickelungsgeschichte des Menschen*. 4th rev. and enlarged ed. Leipzig: Wilhelm Engelmann.

Haeckel, E. (1897). *The evolution of man: A popular exposition of the principle points of human ontogeny and phylogeny*, vols. 1 and 2. New York: D. Appleton.

Haecker, V. (1918). *Entwicklungsgeschichtliche Eigenschaftsanalyse (Phänogenetik)*. Jena, Germany: Gustav Fischer.

Haldane, J. B. S. (1924). A mathematical theory of natural and artificial selection. Part I. *Transactions of the Cambridge Philosophical Society, 23*, 19–41.

Haldane, J. B. S. (1932a). *The causes of evolution*. London: Longmens Green.

Haldane, J. B. S. (1932b). The time of action of genes, and its bearing on some evolutionary problems. *American Naturalist, 56*, 5–24.

Haldane, J. B. S. (1946). The interaction of nature and nurture. *Annals of Human Genetics, London, 13*, 197–205.

Hall, B. K. (1988). The embryonic development of bone. *American Scientist, 76*, 174–181.

Hall, C. S. (1938). The inheritance of emotionality. *Sigma Xi Quarterly, 26*, 17–27.

Hardy, A. C. (1965). *The living stream*. London: Collins.

Harlow, H. F., Dodsworth, R. O., & Harlow, M. K. (1965). Total social isolation in monkeys. *Proceedings of the National Academy of Sciences USA, 54*, 90–96.

Hayashi, Y. (1965). Differentiation of the beak epithelium as studied by a xenoplastic induction system. *Japanese Journal of Experimental Morphology, 19*, 116–123.

Heaton, M. B., & Klein, S. L. (1981). Recovery from experimentally induced problem-solving deficits in neonatal Peking ducklings as a function of environmental stimulation. *Developmental Psychobiology, 14*, 59–65.

Hebb, D. O. (1947). The effects of early experience on problem-solving at maturity. *American Psychologist, 2*, 306–307.

Hebb, D. O. (1949). *The organization of behavior*. New York: Wiley.

Hebb, D. O., & Williams, K. (1946). A method of rating animal intelligence. *Journal of General Psychology, 34*, 59–65.

Heron, W. T. (1935). The inheritance of maze learning ability in rats. *Journal of Comparative Psychology, 19*, 77–89.

Hirsch, J., & Boudreau, J. C. (1958). Studies in experimental behavior genetics. I. The heritability of phototaxis in *Drosophila melanogaster. Journal of Comparative and Physiological Psychology, 51*, 647–651.

Hirsch, J., & Erlenmeyer-Kimmling, L. (1961). Sign of taxis as a property of the genotype. *Science, 134*, 835–836.

His, W. (1888). On the principles of animal morphology. *Proceedings of the Royal Society of Edinburgh, 15*, 287–298.

Ho, M.-W. (1984). Environment and heredity in development and evolution. In M.-W. Ho & P. T. Saunders (Eds.), *Beyond neo-Darwinism: An introduction to the new evolutionary paradigm*. London: Academic Press.

Ho, M.-W. et al. (1983). Effects of successive generations of ether treatment on penetrance and expression of the bithorax phenocopy in *Drosophila melanogaster. Journal of Experimental Zoology, 225*, 357–368.

Ho, M.-W., & Fox, S. W. (Eds.) (1988). *Evolutionary processes and metaphors*. New York: Wiley.

Ho, M.-W., & Saunders, P. T. (1982). The epigenetic approach to the evolution of organisms—With notes on its relevance to social and cultural evolution. In H. C. Plotkin (Ed.), *Learning, development, and culture: Essays in evolutionary epistemology*. London: Wiley.

Ho, M.-W., & Saunders, P. T. (Eds.) (1984). *Beyond neo-Darwinism: An introduction to the new evolutionary paradigm*. London: Academic Press.

Honey, R. C. (1990). Stimulus generalization as a function of stimulus novelty and familiarity in rats. *Journal of Experimental Psychology: Animal Behavior Processes, 16,* 178–184.

Hopson, J. A. (1977). Relative brain size and behavior in archosaurian reptiles. *Annual Review of Ecology and Systematics, 8,* 429–448.

Horn, G. (1985). *Memory, imprinting, and the brain.* Oxford: Clarendon Press.

Horowitz, F. D. (1987). *Exploring developmental theories: Toward a structural/behavioral model of development.* Hillsdale, New Jersey: Erlbaum.

Huck, W. V., Labov, J. B., & Lisk, R. D. (1986). Food restricting young hamsters (*Mesocricetus auratus*) affects sex ratio and growth of subsequent offspring. *Biology of Reproduction, 35,* 592–598.

Hughes, K. R., & Zubek, J. P. (1956). Effect of glutamic acid on the learning ability of bright and dull rats. I. Administration during infancy. *Canadian Journal of Psychology, 10,* 132–138.

Hutchinson, G. E., & Rachootin, S. (1979). Historical introduction. In reprint of W. Bateson's (1913) *Problems of genetics.*

Huxley, J. S. (1942). *Evolution: The modern synthesis.* London: Allen & Unwin. (Reprinted in 1963 by Dover Press, New York.)

Huxley, J. S. (1957). The three types of evolutionary progress. *Nature, 180,* 454–455.

Huxley, T. H. (1870). *Lay sermons, addresses, and reviews.* London: Macmillan.

Hydén, H., & Egyházi, E. (1962). Nuclear RNA changes of nerve cells during a learning experiment in rats. *Proceedings of the National Academy of Science, 48,* 1366–1373.

Hydén, H., & Egyházi, E. (1964). Changes in RNA content and base composition in cortical neurons of rats in a learning experiment involving transfer of handedness. *Proceedings of the National Academy of Science, 52,* 1030–1035.

Hymovitch, B. (1952). The effects of experimental variations on problem solving in the rat. *Journal of Comparative and Physiological Psychology, 45,* 313–321.

Ingham, P. W. (1988). The molecular genetics of embryonic pattern formation in *Drosophila. Nature, 335,* 25–34.

Jaisson, P. (1975). L'impregnation dans l'ontogenese des comportements de soins aux cocons chez la jeune formi rousse (*Formica polyctena* Forst). *Behaviour, 52,* 1–37.

Jerison, H. J. (1968). Brain evolution and *Archaeopteryx. Nature, 219,* 1381–1382.

Jerison, H. J. (1969). Brain evolution and dinosaur brains. *American Naturalist, 103,* 575–588.

Jerison, H. J. (1973). *Evolution of the brain and intelligence.* New York: Academic Press.

Johnston, T. D. (1987). The persistence of dichotomies in the study of behavioral development. *Developmental Review, 7,* 149–182.

Johnston, T. D., & Gottlieb, G. (1990). Neophenogenesis: A developmental theory of phenotypic evolution. *Journal of Theoretical Biology, 147,* 471–495.

Jollos, V. (1934). Inherited changes produced by heat treatment in *Drosophila melanogaster. Genetics, 16,* 476–494.

Kevles, D. J. (1985). *In the name of eugenics: Genetics and the uses of human heredity.* New York: A. A. Knopf.

Kimura, K. (1984). Studies on growth and development in Japan. *Yearbook of Physical Anthropology, 27,* 179–214.

Klineberg, O. (1935). *Negro intelligence and selective migration.* New York: Columbia University Press.

Kollar, E. J., & Fisher, C. (1980). Tooth induction in chick epithelium: Expression of quiescent genes for enamel synthesis. *Science, 207,* 993–995.

Krushinsky, L. V. (1965). Solution of elementary logical problems by animals on the basis of extrapolation. *Progress in Brain Research, 17,* 280–308.

Kuo, Z.-Y. (1976). *The dynamics of behavior development: An epigenetic view* (enlarged ed.). New York: Plenum Press.

Lamarck, J. B. (1984). *Zoological philosophy: An exposition with regard to the natural history of animals.* Chicago: University of Chicago Press. (Originally published in French in 1809; first translated into English in 1914.)

Larson, A., Prager, E. M., & Wilson, A. C. (1984). Chromosomal evolution, speciation and morphological change in vertebrates: The role of social behavior. *Chromosomes Today, 8,* 215–228.

Lehrman, D. S. (1970). Semantic and conceptual issues in the nature-nurture problem. In L. R. Aronson, D. S. Lehrman, E. Tobach, & J. S. Rosenblatt (Eds.), *Development and evolution of behavior.* San Francisco, California: W. H. Freeman.

Leonovicová V., & Novák, V. J. A. (Eds.) (1987). *Behavior as one of the main factors of evolution.* Praha, Czechoslovakia: Czechoslovak Academy of Sciences.

Lerner, I. M. (1958). *The genetic basis of evolution.* New York: Wiley.

Lerner, I. M. (1968). *Heredity, evolution, and society.* San Francisco, California: W. H. Freeman.

Lerner, R. M., & Kaufman, M. B. (1985). The concept of development in contextualism. *Developmental Review, 5,* 309–333.

Levine, S. (1956). A further test of infantile handling and adult avoidance learning. *Journal of Personality, 25,* 70–80.

Levine, S. (1962). The effects of infantile experience on adult behavior. In A. J. Bachrach (Ed.), *Experimental foundations of clinical psychology.* New York: Basic Books.

Lewin, B. (Ed.) (1980). *Gene expression.* Vol. 2. *Eucaryotic chromosomes,* 2nd ed. New York: Wiley.

Lewontin, R. (1974). *Genetic basis of evolutionary change.* New York: Columbia University Press.

Lickliter, R., & Berry, T. D. (1990). The phylogeny fallacy: Developmental psychology's misapplication of evolutionary theory. *Developmental Review, 10,* 348–364.

Løvtrup, S. (1987). *Darwinism: Refutation of a myth.* Beckenham, Kent, England: Croom Helm.

Macphail, E. M., & Reilly, S. (1989). Rapid acquisition of a novelty versus familiarity concept by pigeons (*Columba livia*). *Journal of Experimental Psychology: Animal Behavior Processes, 15,* 242–252.

Magnusson, D. (1988). *Individual development from an interactional perspective: A longitudinal study.* Hillsdale, New Jersey: Erlbaum.

Manning, A. (1961). The effects of artificial selection for mating speed in *Drosophila melanogaster. Animal Behaviour, 9,* 82–92.

Manning, A. (1963). Selection for mating speed in *Drosophila melanogaster* based on the behaviour of one sex. *Animal Behaviour, 11,* 116–120.

Mason, W. A. (1968). Early social deprivation in the nonhuman primates: Implications for human behavior. In D. Glass (Ed.), *Environmental influences.* New York: Rockefeller University Press.

Matsuda, R. (1987). *Animal evolution in changing environments.* New York: Wiley.

Mayr, E. (1942). *Systematics and the origin of species.* New York: Columbia University Press.

Mayr, E. (1954). Change of genetic environment and evolution. In J. Huxley, A. C. Hardy, & E. B. Ford (Eds.), *Evolution as a process.* London: Allen and Unwin.

Mayr, E. (1958). Behavior and systematics. In A. E. Roe, & G. G. Simpson (Eds.), *Behavior and evolution.* New Haven: Yale University.

Mayr, E. (1963). *Animal species and evolution.* Cambridge, Massachusetts: Harvard University Press.

Mayr, E. (1982). *The growth of biological thought.* Cambridge, Massachusetts: Belknap Press of Harvard University Press.

Mayr, E. (1988). *Toward a new philosophy of biology: Observations of an evolutionist.* Cambridge, Massachusetts: Belknap Press of Harvard University Press.

McDonald, P., & Topoff, H. (1985). Social regulation of behavioral development in the ant, *Novomessor albisetosus* (Mayr). *Journal of Comparative Psychology, 99,* 3–14.

Mendel, G. (1966). Experiments on plant hybrids. In C. Stern, & E. R. Sherwood (Eds.), *The origin of genetics: A Mendel sourcebook*. San Francisco, California: Freeman. (Translation of Mendel's German-language article, dated 1865.)

Mirsky, A. E., & Ris, H. (1951). The desoxyribonucleic acid content of animal cells and its evolutionary significance. *Journal of General Physiology, 34,* 451–462.

Mivart, St. G. (1871). *On the genesis of species*. London: Macmillan.

Montagu, M. F. A. (1962). Time, morphology, and neoteny in the evolution of man. In M. F. A. Montagu (Ed.), *Culture and the evolution of man*. New York: Oxford University Press.

Montagu, M. F. A. (1966). Constitutional and prenatal factors in infant and child health. In M. J. Senn (Ed.), *Symposium on the healthy personality*. New York: Josiah Macy Foundation.

Morgan, T. H. (1934). *Embryology and genetics*. Westport, Connecticut: Greenwood.

Myers, M. M., Brunelli, S. A., Shair, H. N., Squire, J. M., & Hofer, M. A. (1989). Relationships between maternal behavior of SHR and WKY dams and adult blood pressures of cross-fostered F_1 pups. *Developmental Psychobiology, 22,* 55–67.

Myers, M. M., Brunelli, S. A., Squire, J. M., Shindeldecker, R. D., & Hofer, M. A. (1989). Maternal behavior of SHR rats and its relationship to offspring blood pressures. *Developmental Psychobiology, 22,* 29–53.

Nachman, M. (1959). The inheritance of saccharin preference. *Journal of Comparative and Physiological Psychology, 52,* 451–457.

Needham, J. (1959). *A history of embryology*. London and New York: Abelard-Schuman.

Newman, S. A. (1988). Idealist biology. *Perspectives in Biology and Medicine, 31,* 353–368.

Oppenheimer, J. M. (1940). The non-specificity of the germ-layers. *Quarterly Review of Biology, 15,* 1–27.

Oppenheimer, J. M. (1967). *Essays in the history of embryology and biology*. Cambridge, Massachusetts: M.I.T. Press.

Oster, G., Odell, G., & Alberch, P. (1980). Mechanics, morphogenesis, and evolution. *Lectures on Mathematics in the Life Sciences, 13,* 165–255.

Oyama, S. (1985). *The ontogeny of information*. Cambridge, England: Cambridge University Press.

Oyama, S. (1989). Transmission and construction: Levels and the problem of heredity. In G. Greenberg & E. Tobach (Eds.), *Levels of social behavior: Evolutionary and genetic aspects*. New York: Gordian Press.

Parsons, P. A. (1981). Habitat selection and speciation in *Drosophila*. In

W. R. Atchley, & D. S. Woodruff (Eds.), *Evolution and speciation: Essays in honor of M. J. D. White*. Cambridge, England: Cambridge University Press.

Pearson, K. (1901). Editorial. *Biometrika, 1,* 1–6.

Piaget, J. (1978). *Behavior and evolution.* New York: Pantheon Books.

Platt, S. A., & Sanislow, C. A. (1988). Norm-of-reaction: Definition and misinterpretation of animal research. *Journal of Comparative Psychology, 102,* 254–261.

Plomin, R. (1986). *Development, genetics, and psychology.* Hillsdale, New Jersey: Erlbaum.

Plotkin, H. C. (Ed.) (1988). *The role of behavior in evolution.* Cambridge, Massachusetts: M.I.T. Press.

Pritchard, D. J. (1986). *Foundations of developmental genetics.* London and Philadelphia: Taylor and Francis.

Provine, W. B. (1971). *The origin of theoretical population genetics.* Chicago: University of Chicago Press.

Provine, W. B. (1986). *Sewall Wright and evolutionary biology.* Chicago: University of Chicago Press.

Rabinovitch, M. S., & Rosvold, H. E. (1951). A closed-field intelligence test for rats. *Canadian Journal of Psychology, 5,* 122–128.

Raff, R. A., & Kaufman, T. C. (1983). *Embryos, genes, and evolution.* New York: Macmillan.

Razran, G. (1971). *Mind in evolution.* New York: Houghton Mifflin.

Reid, R. G. B. (1985). *Evolutionary theory: The unfinished synthesis.* Ithaca, New York: Cornell University Press.

Renner, M. J., & Rosenzweig, M. R. (1987). *Enriched and impoverished environments.* New York: Springer.

Rensch, B. (1959). *Evolution above the species level.* New York: Columbia University Press.

Richards, R. J. (1987). *Darwin and the emergence of evolutionary theories of mind and behavior.* Chicago: University of Chicago Press.

Rodgers, D. A., & McClearn, G. E. (1962). Alcohol preference in mice. In E. L. Bliss (Ed.), *Roots of behavior.* New York: P. B. Hoeber.

Rogers, L. J. (1990). Light input and the reversal of functional lateralization in the chicken brain. *Behavioural Brain Research, 38,* 211–221.

Rosen, D. E., & Buth, D. G. (1980). Empirical evolutionary research versus neo-Darwinian speculation. *Systematic Zoology, 29,* 300–308.

Rosenzweig, M. R., & Bennett, E. L. (1978). Experiential influences on brain anatomy and brain chemistry in rodents. In G. Gottlieb (Ed.), *Early influences.* New York: Academic Press.

Rothenbuhler, W. C. (1967). Genetic and evolutionary considerations of social behavior of honeybees and some related insects. In J. Hirsch (Ed.), *Behavior-genetic analysis.* New York: McGraw-Hill.

Roux, W. (1888/1974). Contributions to the developmental mechanics of the embryo. On the artificial production of half-embryos by destruction of one of the first two blastomeres, and the later development (postgeneration) of the missing half of the body. (Translated from the German in 1974 in B. H. Willier & J. M. Oppenheim [Eds.], *Foundations of experimental embryology.* New York: Hafner.)

Roux, W. (1894). The problems, methods and scope of developmental mechanics. An introduction to the "Archiv für Entwicklungsmechanik der Organismen," translated by W. M. Wheeler. In *Biological Lectures of the Marine Biological Laboratory of Woods Hole, Mass., 1895.* Boston: Ginn & Co., pp. 149–190.

Rundquist, E. A. (1933). Inheritance of spontaneous activity in rats. *Journal of Comparative Psychology, 16,* 415–438.

Sapp, J. (1987). *Beyond the gene: Cytoplasmic inheritance and the struggle for authority in genetics.* New York: Oxford University Press.

Sackett, G. P. (1968). Abnormal behavior in laboratory-reared rhesus monkeys. In M. W. Fox (Ed.), *Abnormal behavior in animals.* Philadelphia: W. B. Saunders.

Sameroff, A. J. (1983). Developmental systems: Contexts and evolution. In P. H. Mussen (Ed.), *Handbook of child psychology* (vol. 1): W. Kessen (Ed.), *History, theory, and methods.* New York: Wiley.

Sarich, V. (1980). A macromolecular perspective on the material basis of evolution. In L. K. Piternick (Ed.), *Richard Goldschmidt: Controversial geneticist and creative biologist.* Boston: Birkhäuser.

Schmalhausen, I. I. (1949). *Factors of evolution: The theory of stabilizing selection.* Philadelphia: Blakiston.

Schneirla, T. C. (1961). Instinctive behavior, maturation—experience and development. In B. Kaplan & S. Wapner (Eds.), *Perspectives in psychological theory—Essays in honor of Heinz Werner.* New York: International Universities Press.

Searle, L. V. (1949). The organization of hereditary maze-brightness and maze-dullness. *Genetic Psychology Monographs, 39,* 279–325.

Sewertzoff, A. N. (1929). Directions of evolution. *Acta Zoologica* (Stockholm), *10,* 59–141.

Shapiro, D. Y. (1981). Serial female sex changes after simultaneous removal of males from social groups of a coral reef fish. *Science, 209,* 1136–1137.

Shatz, C. J., & Stryker, M. P. (1988). Prenatal tetrodotoxin infusion blocks segregation of retinogeniculate afferents. *Science, 242,* 87–89

Shields, J. (1962). *Monozygotic twins.* London: Oxford University Press.

Simpson, G. G. (1944). *Tempo and mode in evolution.* New York: Columbia University Press. (Reissued, with a new introduction, in 1984.)

Simpson, G. G. (1953). The Baldwin effect. *Evolution, 7,* 110–117.

Singh, A. K., & Lakhotin, S. C. (1988). Effect of low-temperature rearing on heat shock protein synthesis and heat sensitivity in *Drosophila melanogaster. Developmental Genetics, 9,* 193–201.

Singh, R. S. (1989). Population genetics and the evolution of species related to *Drosophila melanogaster. Annual Review of Genetics, 23,* 425–453.

Skolnick, N., Ackerman, S., Hofer, M. A., & Weiner, H. (1980). Vertical transmission of acquired ulcer susceptibility in the rat. *Science, 208,* 1161–1163.

Sparrow, A. H., Price, H. J., & Underbrink, A. G. (1972). A survey of DNA content per cell and per chromosome of prokaryotic and eukaryotic organisms: Some evolutionary considerations. *Brookhaven Symposium on Biology, 23,* 451–494.

Spencer, H. (1898). *The principles of psychology,* vol. 1, 3rd ed. New York: D. Appleton.

Stanley, S. M. (1981). *The new evolutionary timetable: Fossils, genes, and the origin of species.* New York: Basic Books.

Stebbins, G. L. (1950). *Variation and evolution in plants.* New York: Columbia University Press.

Steele, E. J. (1979). *Somatic selection and adaptive evolution.* Toronto: Williams and Wallace.

Tanner, J. M. (1978). *Foetus into man: Physical growth from conception to maturity.* Cambridge: Harvard University Press.

Thomas, C. A. (1971). The genetic organization of chromosomes. *Annual Review of Genetics, 5,* 237–256.

Thompson, W. D'Arcy (1917). *On growth and form.* Cambridge, England: Cambridge University Press.

Tolman, E. C. (1924). The inheritance of maze-learning ability in rats. *Journal of Comparative Psychology, 4,* 1–18.

Trivers, R. (1985). *Social evolution.* Menlo Park, California: Benjamin-Cummings.

Tryon, R. (1942). Individual differences. In F. A. Moss (Ed.), *Comparative psychology.* New York: Prentice-Hall.

Uphouse, L. L., & Bonner, J. (1975). Preliminary evidence for the effects of environmental complexity on hybridization of rat brain RNA to rat unique DNA. *Developmental Psychobiology, 8,* 171–178.

Valsiner, J. (1987). *Culture and the development of children's action.* Chichester, England: Wiley.

Waddington, C. H. (1956). Genetic assimilation of the *bithorax* phenotype. *Evolution, 10,* 1–13.

Waddington, C. H. (1959). Canalization of development and genetic assimilation of acquired characters. *Nature, 183,* 1654–1655.

Waddington, C. H. (1961). Genetic assimilation. *Advances in Genetics, 10,* 257–293.

Waddington, C. H. (1969). The theory of evolution today. In A. Koestler, & J. R. Smythies (Eds.), *Beyond reductionism.* London: Hutchinson.

Waddington, C. H. (1975). *The evolution of an evolutionist.* New York: Cornell University Press.

Wallman, J. (1979). A minimal visual restriction experiment: Preventing chicks from seeing their feet affects later responses to mealworms. *Developmental Psychobiology, 12,* 391–397.

Weismann, A. (1894). *The effect of external influences upon development.* London: Henry Frowde.

Weiss, P. (1959). Cellular dynamics. *Reviews of Modern Physics, 31,* 11–20.

Weldon, W. F. R. (1901). Professor de Vries on the origin of species. *Biometrika, 1,* 365–374.

Wessells, N. K. (1977). *Tissue interactions and development.* Menlo Park, California: W. A. Benjamin.

West, M. J., & King, A. P. (1987). Settling nature and nuture into an ontogenetic niche. *Developmental Psychobiology, 20,* 549–562.

Whetstone, K. N. (1983). Braincase of mesozoic birds: I. New preparation of the "London" *Archaeopteryx. Journal of Vertebrate Paleontology, 2,* 439–452.

Wigglesworth, V. B. (1964). *The life of insects.* Cleveland, Ohio: World Publishing Co.

Williams, G. C. (1982). A comment on genetic assimilation. In H. C. Plotkin (Ed.), *Learning, development, and culture: Essays in evolutionary epistemology.* London: Wiley.

Wilson, A. C. (1985). The molecular basis of evolution. *Scientific American, 253,* 164–173.

Wolpert, L. (1971). Positional information and pattern formation. *Current Topics in Developmental Biology, 6,* 183–224.

Wolpert, L., & Stein, W. D. (1982). Evolution and development. In H. C. Plotkin (Ed.), *Learning, development, and culture: Essays in evolutionary epistemology.* London: Wiley.

Wright, S. (1920). The relative importance of heredity and environment in determining the piebald pattern of guinea pigs. *Proceedings of the National Academy of Sciences USA, 6,* 320–332.

Wright, S. (1922). The effects of inbreeding and crossbreeding on guinea pigs. *Bulletin of the U.S. Department of Agriculture,* nos. 1090, 1121.

Wright, S. (1968). *Evolution and the genetics of populations,* vol. 1: *Genetic and biometric foundations.* Chicago: University of Chicago Press.

Wright, S. (1980). Genic and organismic selection. *Evolution, 34,* 825–843.

Wyles, J. S., Kunkel, J. G., & Wilson, A. C. (1983). Birds, behavior, and anatomical evolution. *Proceedings of the National Academy of Sciences USA, 80,* 4394–4397.

Yule, G. U. (1902). Mendel's laws and their probable relations to intra-racial heredity. *New Phytologist, 1,* 193–207, 222–238.

Zamenhof, S., & van Marthens, E. (1978). Nutritional influences on prenatal brain development. In G. Gottlieb (Ed.), *Early influences.* New York: Academic Press.

Zamenhof, S., & van Marthens, E. (1979). Brain weight, brain chemical content, and their early manipulation. In M. E. Hahn, C. Jensen, & B. C. Dudek (Eds.), *Development and evolution of brain size.* New York: Academic Press.

Zirkle, C. (1946). The early history of the idea of the inheritance of acquired characters and of pangenesis. *Transactions of the American Philosophical Society, 35,* 91–150.

Index